MODERN CONCEPTS IN CHEMISTRY

EDITORS

Bryce Crawford, Jr., University of Minnesota
W. D. McElroy, Johns Hopkins University
Charles C. Price, University of Pennsylvania

BOOKS PUBLISHED

MICHAEL J. S. DEWAR, University of Chicago—*Hyperconjugation*

JACK HINE, Georgia Institute of Technology—*Divalent Carbon*

LARS MELANDER, Nobel Institute of Chemistry, Stockholm—*Isotope Effects on Reaction Rates*

BOOKS IN PREPARATION

B. E. CONWAY, University of Ottawa—Theory of Electrode Processes

R. CLINTON FULLER, Dartmouth Medical School—Photosynthesis

PHILIP GEORGE and ROBERT J. RUTMAN, University of Pennsylvania—Thermodynamic Driving Force in Biosynthetic Reactions

HAROLD HART, Michigan State University—Carbonium Ions in Organic Reactions

HAROLD S. JOHNSTON, University of California, Berkeley—Gas Phase Reaction Rate Theory

MICHAEL KASHA, Florida State University—Molecular Excitation

JOHN L. MARGRAVE, Rice University—High Temperature Chemistry

PATRICK A. McCUSKER, University of Notre Dame—Organo-Boron Compounds

E. E. MUSCHLITZ, JR. and THOMAS L. BAILEY, University of Florida—The Mass Spectrometer as a Research Instrument

C. G. OVERBERGER, Polytechnic Institute of Brooklyn; JOSEPH G. LOMBARDINO, Charles Pfizer and Co., Inc.; and JEAN-PIERRE ANSELME, Polytechnic Institute of Brooklyn—Chemistry of Organic Compounds with Nitrogen-Nitrogen Bonds

BERNARD PULLMAN and ALBERTE PULLMAN, University of Paris—Chemical Carcinogenesis in Molecular and Quantum Biology

MICHAEL SZWARC, State University of New York College of Forestry at Syracuse University:
  Free Radicals: Their Formation and Disappearance
  Reactions of Free Radicals: Addition and Atom Abstraction

HOWARD G. TENNENT, Hercules Powder Co.—Organometallic Polymerization

EDWARD R. THORNTON, University of Pennsylvania—Solvolysis Mechanisms

JOHN S. WAUGH, Massachusetts Institute of Technology—Analysis of Nuclear Resonance Spectra

JOHN E. WERTZ, University of Minnesota—Electron Spin Resonance

DIVALENT CARBON

JACK HINE

GEORGIA INSTITUTE OF TECHNOLOGY

THE RONALD PRESS COMPANY · NEW YORK

Library of Congress Catalog Card Number: 64–11755
PRINTED IN THE UNITED STATES OF AMERICA

Preface

This book is a survey of the chemistry of reaction intermediates containing divalent carbon. For this reason the chemistry of stable divalent carbon compounds, such as carbon monoxide and isocyanides, is touched on only lightly. Although many workers in recent years have used the name "carbenes" for divalent carbon derivatives, following the suggestion of Doering (1), we shall follow the *Chemical Abstracts* practice of naming such species as derivatives of methylene, the parent member of the series.

The basis for the understanding of the chemistry of methylenes is to be sought in the study of the mechanisms of the reactions involving methylenes and the effect of structure on reactivity in these reactions. Therefore it is these aspects of the subject that are emphasized in the present work.

In selecting material I have attempted to examine all the relevant literature available to me by July, 1962. Many of the literature citations not listed, especially the older ones, may be found in the references that are listed. It is not implied that the references given or the workers mentioned in connection with given subjects were the *first* to describe evidence on that subject. The assignment of priorities in scientific discoveries is a task the author is ordinarily willing to leave to others.

I should like to acknowledge my indebtedness to Mr. Robert L. McDaniel for his comments on several chapters, to Dr. Benjamin F. Plummer for reading and criticizing the entire manuscript, and to my wife for typing, proofreading, and assisting in all the other tasks required to bring this book into being.

JACK HINE

Atlanta, Georgia
January, 1964

Contents

DIVALENT CARBON

1

Introduction and General References

ABNORMAL VALENCE STATES OF CARBON

The most important reaction intermediates in organic chemistry are probably those containing carbon in an abnormal valence state. Only four types of such intermediates are common. Three types, carbonium ions, carbanions, and free radicals, are trivalent-carbon derivatives in which the trivalent carbon atom bears a positive charge, a negative charge, and no charge (but an unpaired electron), respectively.*

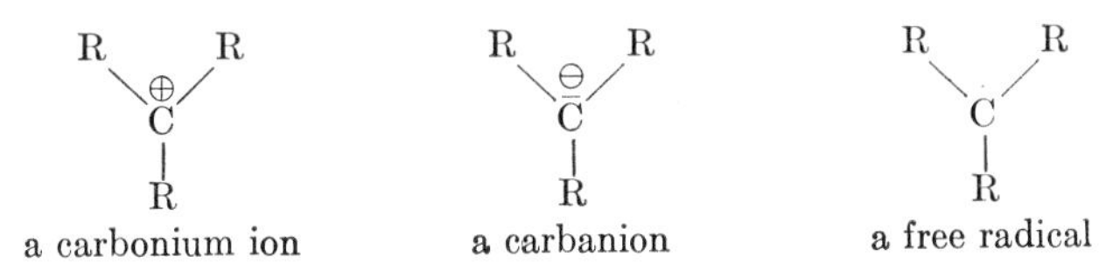

Methylenes, uncharged divalent-carbon derivatives, comprise the fourth type, which may be subdivided into singlets (containing no unpaired electrons) and triplets (containing two unpaired electrons).

* In the formulas in this book an electron pair will be represented by a straight line, between two atomic symbols when the electron pair is bonding the two atoms, and parallel to the side of an atomic symbol when representing an unshared electron pair. Unpaired electrons will be represented by dots.

Methylenes

$R—\overline{C}—R$	$R—\dot{\underset{\cdot}{C}}—R$
Singlet (non-radical)	Triplet (diradical)

Both types are important and will be discussed in this book.

HISTORY OF METHYLENES

In spite of the rapid recent growth of interest in methylenes the subject is not a new one. Dumas and Péligot, in 1835, suggested that pure methylene could be made by the pyrolysis of methyl chloride at the proper temperature.* Later, Butlerov obtained ethylene from the reaction of methylene iodide with copper and made the plausible suggestion that it arose by the dimerization of methylene. The proposal of Geuther, made in 1862, that the basic hydrolysis of chloroform involves the intermediate formation of dichloromethylene has now been well established. Around the turn of the century, Nef published a series of articles in which a wide variety of organic reactions were said to proceed via the intermediate formation of methylenes. Although some of Nef's suggested reaction mechanisms appear reasonable in the light of modern theory, many others are clearly incorrect, and even for the reasonable mechanisms little strong evidence was offered.

Modern work in the field of methylenes may be said to have begun around 1910 with the investigations of Staudinger, whose experimental observations are of great significance and whose mechanistic interpretations require little editing by the modern chemist. It was not until about 1950, however, that the recent phenomenal growth in interest in divalent-carbon intermediates began. A literature search reveals that between 1949 and 1962 the number of papers published annually on the subject increased more than tenfold.

* It should be noted that since the atomic weight of 12 for carbon was not generally accepted until after 1862, in some of the early suggestions methylene was given such formulas as C_2H_2 instead of CH_2.

GENERAL REFERENCES

Knunyants, Gambaryan, and Rokhlin have reviewed (in Russian) the literature on methylenes through early 1958 in detail (2). Shorter reviews were published by Kirmse in 1959 and 1961 (3, 4), by Miginiac in 1962 (4a), and by Chinoporos in 1963 (4b). Huisgen's review on diazo compounds (5) contains much information about methylenes, as does Zollinger's *Azo and Diazo Chemistry, Aliphatic and Aromatic Compounds* (6). Parham and Schweizer reviewed work on halomethylenes with emphasis on addition to olefins (6b).

Reaction mechanisms and reactivity have been emphasized in the present author's discussion of methylenes in a textbook of physical organic chemistry (7).

TYPES OF MECHANISMS FOR THE FORMATION AND REACTIONS OF METHYLENES

A number of kinds of reaction mechanisms can be envisaged for the formation of methylenes. These include the following types (among others).

1. The homolysis* of a double bond, e.g.,†

$$\begin{matrix} R' \\ \quad\diagdown \\ \quad\quad C{=}X \rightarrow R{-}C{-}R' + X \\ \quad\diagup \\ R \end{matrix}$$

2. A carbanion may lose‡ a group with its bonding electron pair, e.g.,

$$R{-}\overset{\ominus}{\underset{\displaystyle R'}{\underset{|}{\bar{C}}}}{-}X \rightarrow R{-}C{-}R' + X^-$$

* In the homolysis of a bond, the bonding electrons are divided equally between the resulting fragments; in heterolysis they are divided unequally.

† None of the non-bonding outer electrons are shown in the methylenes produced in the following equations in order to leave open the possibility of the formation of either a singlet or a triplet.

‡ We shall define loss of a group in this classification of mechanisms to include removal of the group by the attack of some reagent as well as spontaneous loss of the group.

3. A carbonium ion may lose a group without its bonding electron pair, e.g.,

$$\mathrm{R{-}\overset{\oplus}{C}(R'){-}X \rightarrow R{-}C{-}R' + X^{+}}$$

4. A free radical may lose a group with one of its bonding electrons, e.g.,

$$\mathrm{R{-}\dot{C}(R'){-}X \rightarrow R{-}C{-}R' + X\cdot}$$

5. Two groups attached to a tetravalent carbon atom may simultaneously form a bond to each other and break their bonds to carbon, e.g.,

$$\mathrm{RR'CXY \rightarrow R{-}C{-}R' + X{-}Y}$$

6. Two groups attached to a tetravalent carbon atom may be lost simultaneously in some other way, e.g.,

$$\mathrm{RR'CXY \rightarrow R{-}C{-}R' + X^{+} + Y^{-}}$$

7. An unsaturated carbanion, free radical, or carbonium ion may lose a γ-substituent group with two, one, or none of its bonding electron pairs to give a species for which a methylenic contributing structure may be written, e.g.,

$$\mathrm{{-}C(X){-}C{=}\overset{\ominus}{C}{-} \rightarrow \left[{-}\overset{\oplus}{C}{-}C{=}\overset{\ominus}{C}{-} \leftrightarrow {-}C{=}C{-}C{-} \right] + X^{-}}$$

The formation of methylene from diazomethane or ketene (Chapter 2) is a reaction of type 1; the formation of dichloromethylene from chloroform can be accomplished by reactions of type 2 or type 5, depending on the conditions (Chapter 3); the reaction of methylene bromide with sodium in the vapor phase (Chapter 2) is a type 4 reaction; chlorodifluoromethane reacts with base by a mechanism of type 6 (Chapter 3); and the 1,3-dehydrohalogenation of 3-chloro-3-methyl-1-butyne (Chapter 7) is a reaction of type 7. The deprotonation of a thiazolium salt (Chapter 8) is a reaction of type 3.

Each of the types of mechanisms for formation of methylenes, when written in reverse, becomes a mechanism for the reaction of a methylene. Some specific examples are:

1. $CH_2 + CO \rightarrow CH_2{=}CO$
2. $CCl_2 + I^- \rightarrow CCl_2I^-$
3. $R_2N{-}\overline{C}{-}NR_2 \xrightarrow{H^+} HC(NR_2)_2{}^+$
4. $H{-}\dot{\underset{\cdot}{C}}{-}H + C_3H_8 \rightarrow CH_3\cdot + C_3H_7\cdot$
5. $H{-}\overline{C}{-}H + C_2H_6 \rightarrow C_3H_8$
   (The one-step addition of methylene to olefins could also be considered a reaction of this type.)
6. $CF_2 + H_2O + Cl^- \rightarrow CHF_2Cl + OH^-$
7. $(CH_3)_2C{=}C{=}C| + OH^- \rightarrow CH_3{-}C(CH_3)(OH){-}C{\equiv}C|^{\ominus}$

In addition methylenes undergo internal-rearrangement reactions, such as

$$CH_3{-}C{-}H \rightarrow CH_2{=}CH_2$$

The reverse of such a reaction would be a method of generating methylenes but no such method appears to have been established.

Each of the types of reaction mechanism will be discussed in more detail in subsequent chapters.

2
Formation and Reactions of Methylene

Several of the general methods that may be used for the generation of methylenes (cf. Chapter 1) have been used in generating the simplest member of the family, methylene.

In producing methylene by the homolysis of a double bond,

$$CH_2{=}X \rightarrow CH_2 + X$$

the ease of reaction depends on the stability of the other fragment (X) being formed. Because of the considerable stability of nitrogen and carbon monoxide, methylene is often generated by the pyrolysis and photolysis of diazomethane and of ketene.

The structure of the diazomethane produced by such standard methods as the alkaline decomposition of methylnitrosourea was a subject of argument among early workers, some of whom proposed a cyclic structure and others a structure in which the carbon and two nitrogen atoms are arranged in a straight line.

$$\begin{matrix} H \\ & \diagdown \\ & & C{=}\overset{\oplus}{N}{=}\overset{\ominus}{\overline{N}}| \\ & \diagup \\ H \end{matrix} \leftrightarrow \begin{matrix} H \\ & \diagdown \\ & & \overset{\ominus}{\overline{C}}{-}\overset{\oplus}{N}{\equiv}N| \\ & \diagup \\ H \end{matrix} \leftrightarrow \text{etc.}$$

Subsequent work proved that the linear structure is correct. More recently, however, several investigators have used different methods to prepare the isomeric cyclic form of diazomethane.

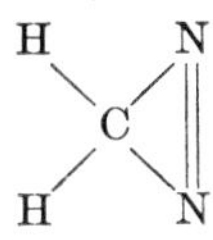

Since the cyclic form of diazomethane has not been widely studied we shall refer to it as cyclodiazomethane or diazirine and reserve the simple term "diazomethane" for the linear form.

FORMATION OF METHYLENE FROM DIAZOMETHANE

Staudinger and Kupfer described the first strong evidence for the formation of free methylene in the pyrolysis of diazomethane in their report that in the presence of carbon monoxide ketene is formed (8):

$$CH_2{=}N_2 \rightarrow CH_2 + N_2$$
$$CH_2 + CO \rightarrow CH_2{=}CO$$

They did not, however, disprove the possibility that the ketene arose from a direct reaction between carbon monoxide and diazomethane without the intermediacy of methylene.

Removal of Metal Mirrors by Methylene. Rice and Glasebrook obtained convincing evidence for the existence of methylene (9) by use of a modification of the technique that had been devised earlier by Paneth for the detection of free radicals. Diazomethane (in ether or butane) was passed at low pressure through a quartz tube that was heated to 400–600° at a given point. It was then found that various

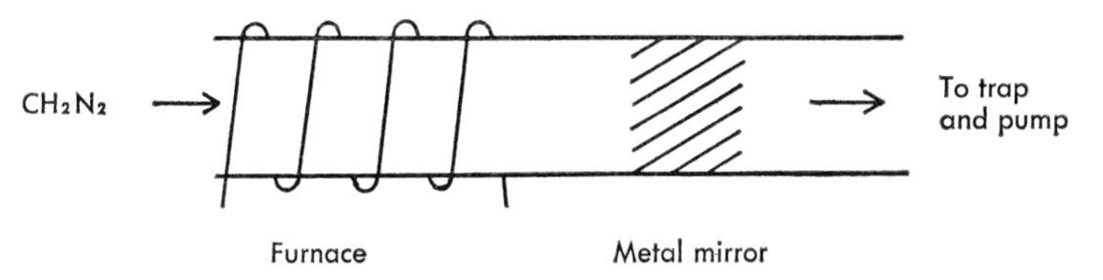

metal mirrors (tellurium, selenium, antimony, and arsenic) that had previously been deposited on the inside of the tube somewhat downstream from the point of heating were removed. The metal must have been removed by the attack of reactive free methylene formed by the decomposition of diazomethane. The mirror-removing agent was not a simple organic radical like methyl or ethyl; it was incapable of removing zinc, cadmium, lead, and other metal mirrors that methyl and ethyl radicals remove quite effectively. Furthermore, the product of removal of tellurium mirrors was shown to be polytelluroformaldehyde, $(CH_2Te)_n$. Subsequent work showed that the rate of removal of metal mirrors decreases as the distance between the mirror and the heated section of the tube increases, and when this distance is large enough all the intermediate methylene dimerizes or undergoes some other reaction before it reaches the mirror, which then is not removed at all (10). The *photolysis* of diazomethane was also shown to produce a fragment capable of removing selenium and tellurium mirrors.

Direct Measurements on Methylene. The most direct and most convincing evidence for the formation of methylene comes from the work of Herzberg and Shoosmith, who studied the flash photolysis of diazomethane diluted with nitrogen (11). New absorption lines, not due to such known species as CH_2N_2, N_2, CH_3, and CH, were found in the vacuum ultraviolet around 1400 Å and in the visible and near infrared between 5500 and 9500 Å. The shift observed when CD_2N_2 was photolyzed shows that the spectrum was that of a hydrogen-containing species, and the fact that three, and only three, sets of lines (one being the set obtained from CH_2N_2, another that from CD_2N_2, and the third set a new one) were obtained when partially deuteriated diazomethane was photolyzed shows that this species contains two symmetrically located hydrogen atoms, and may thus exist in the XH_2, XD_2, and XDH forms. The appearance of one new set of lines in the photolysis of diazomethane containing 60 per cent ^{13}C shows that the species contains one and only one carbon atom. The structure of some of the individual absorption bands rules out the possibility that they are due to any more complicated species (such as an excited state of diazomethane) than CH_2.

The absorption in the range 5500–9500 Å is strongest when the ratio of nitrogen to diazomethane is between 50-to-1 and 100-to-1, while the strength of absorption around 1400 Å increases steadily as the ratio of nitrogen to diazomethane increases, up to 500-to-1 at least. Since both types of absorption were shown to be due to methylene there must be two forms of methylene present. Detailed examination of the spectra showed that the species absorbing around 1400 Å is a triplet form of methylene with a linear structure and C—H bond distance of about 1.03 Å, and that the species absorbing in the 5500–9500 Å range is a singlet with an H—C—H bond angle of about 103° and C—H bond distances of about 1.12 Å. The photolysis reaction apparently leads first to the formation of singlet methylene containing a large amount of excess energy which would bring about the further decomposition of the methylene if it were not quickly lost by collision with the nitrogen molecules used as a diluent. The singlet may subsequently decompose to the more stable triplet form of methylene. Unfortunately the difference in energy content between singlet and triplet methylene is not known.

In advance of the direct studies on methylene just described, a number of semi-quantum mechanical calculations had been carried out with the aim of predicting whether the ground state of methylene would be a singlet or a triplet. As is so often the case with such calculations, both answers had been obtained several times. The answer that had been obtained more frequently was the wrong one. Most of the calculations did agree that the lowest triplet state should be linear and the lowest singlet state bent. This result follows from the simple argument that any unfilled orbitals will be the higher-lying *p* orbitals. Thus triplet methylene will have one electron in each of two *p* orbitals and the remaining *p* orbital will be hybridized with the *s* orbital to give two collinear *sp* bonding orbitals. (The typical *sp* C—H distance of acetylene is 1.06 Å.) The empty orbital of singlet methylene must be *p* in character. If the unshared electron pair is in an *s* orbital the two bonding orbitals should be *p* and the H—C—H bond angle 90°. If the three occupied orbitals combine to give sp^2 hybridization the bond angle will be 120°. The observed bond angle is about halfway between these values, and being smaller than the angle in sp^3 hybridized compounds like methane, the bonding orbitals are more than 75 per cent *p* in character. It is therefore not surprising that the C—H bond distance is somewhat larger than that in methane (1.09 Å).

Attempts have been made to trap free methylene in a glassy matrix at very low temperatures, but it is not yet clear whether any of the attempts were successful.

FORMATION OF METHYLENE FROM KETENE

In 1938 Pearson, Purcell, and Saigh summarized the evidence for the formation of methylene in the photolysis of ketene and, in addition, observed that the photolysis produces reactive intermediates capable of removing tellurium and selenium mirrors with the formation of $(CH_2Te)_n$ and $(CH_2Se)_n$, respectively (10). The reaction mechanism

$$CH_2{=}CO \xrightarrow{h\nu} CH_2 + CO$$
$$CH_2 + CH_2{=}CO \rightarrow CH_2{=}CH_2 + CO$$

is supported by the observation that the photolysis of pure ketene by 2654 Å light gives essentially two molecules of carbon monoxide per molecule of ethylene, with quantum yields of 2.0 and 1.0, respectively (12), but that by addition of a sufficient quantity of some reagent (such as ethylene) that combines efficiently with the intermediate methylene, the rate of formation of carbon monoxide may be halved (13).

Ketene does not decompose immediately upon absorption of the incident light; instead, as is typical in photocatalyzed reactions, it is transformed to an excited state which then decomposes at a rate that depends on how much energy it possesses beyond the minimum required for decomposition. With 2654 Å light the amount of excess energy present is so great that the excited ketene molecules decompose faster than they lose their excess energy by collision even at one-atmosphere pressure, and the quantum yield is pressure-independent (in this range at least). When the less energetic 3660 Å light is used, the quantum yield is much lower at one atmosphere, but it increases at lower pressures where the loss of excess energy by collision competes less effectively with the decomposition process (12).

FORMATION OF METHYLENE FROM METHANE

On the basis of a kinetic study of the pyrolysis of methane (at temperatures around 1000°) Kassel suggested that the first step of the reaction consists of the formation of methylene and a molecule of hydrogen (14).

$$CH_4 \rightarrow CH_2 + H_2$$

The kinetic study was, however, complicated by the existence of an induction period, diffusion of hydrogen through the quartz walls of the reaction vessel, possible heterogeneity, and other factors. Belchetz obtained evidence for the formation of methylene in methane pyrolysis by use of metallic mirrors (15), but Rice and Dooley obtained contradictory results (16).

The mechanism of the pyrolysis of methane does not seem to have been firmly established even yet, but added information on the subject may be obtained by consideration of the decomposition of methane by other means. Gevantman and Williams found that the radiolysis (by X-rays or electrons) of methane in the presence of about 0.1 per cent iodine led to the formation of methylene iodide in yields far too great to have arisen from the subsequent iodination of the methyl iodide also formed (17). Letort and Duval passed methane at room temperature through a high-voltage electric discharge just upstream from a tellurium mirror and obtained polytelluroformaldehyde, $(CH_2Te)_n$, but no methyl derivatives of tellurium (18). Wiener and Burton found that an electric discharge in a mixture of methane and deuterium brought about the formation of considerable CH_2D_2 but relatively little of any other deuteriated methanes. This observation could be interpreted in terms of the formation of methylene and its capture by the deuterium present. However, there is a strong possibility that radiolysis and electric-discharge-induced reactions involve the formation of electrically charged intermediates, and therefore it is not easy to prove whether neutral methylene was formed or not (cf. 18a).

The formation of charged intermediates is less probable in the vacuum-ultraviolet photolysis of methane and therefore Mahan and Mandal's observation that the photolysis of a mixture of CH_4 and CD_4 yields largely H_2 and D_2 and relatively little HD provided good evidence that methylene is generated as an intermediate (19).

Strong evidence for a one-step addition of hydrogen to methylene to give methane is provided by the observation of Chanmugam and Burton that the photolysis of ketene at room temperature in the presence of mixtures of hydrogen and deuterium yields CH_4 and CH_2D_2 but practically none of any of the other deuteriated methanes (20).

$$CH_2{=}CO \rightarrow CH_2 + CO$$
$$CH_2 + H_2 \rightarrow CH_4$$
$$CH_2 + D_2 \rightarrow CH_2D_2$$

Bell and Kistiakowsky also obtained evidence for the one-step addition of hydrogen to methylene produced by the photolysis of diazomethane, and they showed that the addition was not appreciably reversible (21). The D_2 present after the reaction was only very slightly diluted by H_2 and HD. Decomposition of the excited methane to hydrogen atoms and methyl radicals did occur, however. This conclusion was supported by the formation of ethane (by dimerization of the methyl radicals) as well as by other evidence.

$$CH_2 + H_2 \rightarrow CH_4{}^*$$
$$CH_4{}^* \rightarrow CH_3\cdot + H\cdot$$
$$2CH_3\cdot \rightarrow CH_3CH_3$$

The formation of ethane was almost completely inhibited by small amounts of oxygen, which is known to be a very effective reagent at capturing methyl radicals (but not methylene). Thus the excited methane molecules in these experiments decomposed largely to methyl radicals and hydrogen atoms, whereas those produced by 1236 Å radiation (containing 230 kcal./mole energy), in the experiments of Mahan and Mandal (19), decomposed largely to methylene and hydrogen. The vacuum ultraviolet-excited methane contained much more excess energy and it was introduced into the molecule in a different manner.

Bell and Kistiakowsky combined their own observations with earlier data to obtain the following values for the energy required for the successive fission of the four bonds of methane.

$$\begin{aligned} D(CH_3{-}H) &= 102 \text{ kcal./mole} \\ D(CH_2{-}H) &= 105 \pm 3 \\ D(CH{-}H) &= 108 \pm 3 \\ D(C{-}H) &= 82.7 \end{aligned}$$

These values correspond to a heat of formation of methylene of about 85 kcal./mole, but significant uncertainties remain. It is not known, for example, how much excess energy was present in the methylene molecules on which some of the measurements were made. Prophet has recently estimated that the heat of formation of methylene is 95 $\pm$5 kcal./mole (21a).

INSERTION OF METHYLENE INTO C—H BONDS

One of the most common reactions of methylene consists of the insertion of the CH_2 group into a single bond.

$$CH_2 + A{-}B \rightarrow A{-}CH_2{-}B$$

The reaction was discovered by Meerwein, Rathjen, and Werner who observed that the photolysis of diazomethane in diethyl ether yields ethyl *n*-propyl ether and ethyl isopropyl ether and in isopropyl alcohol yields all the products that could arise from insertion into C—H and O—H bonds (22).

$$CH_3CHOHCH_3 + CH_2 \rightarrow \begin{matrix} (CH_3)_3COH \\ CH_3CH_2CHOHCH_3 \\ (CH_3)_2CHOCH_3 \end{matrix}$$

The reaction of methylene with molecular hydrogen described in the preceding section is one example of an insertion reaction.

Mechanisms of Methylene Insertions. Insertions into C—H bonds have been found to occur by two different mechanisms that often operate concurrently. In one mechanism the insertion is a direct one-step reaction with a transition state that may be represented as

$$\gt C\text{-------}H \quad \text{(dotted bonds to } CH_2\text{)}$$

while in the other mechanism two intermediate radicals are formed and then combine.

$$RH + CH_2 \rightarrow R\cdot + CH_3\cdot \rightarrow R{-}CH_3$$

Evidence for the free radical mechanism was obtained by Frey and Kistiakowsky, who found that ethane is a by-product in several gas-phase methylations but that its formation is inhibited by oxygen, a reagent known to be effective at capturing intermediate free radicals (23). Even more convincingly, Frey found that all the products shown below were formed in the reaction of methylene with propane (24).

$$CH_3CH_2CH_3 + CH_2 \nearrow CH_3CH_2CH_2\cdot + CH_3\cdot \quad \searrow CH_3\dot{C}HCH_3 + CH_3\cdot$$

$$CH_3CH_2CH_2\cdot + CH_3\cdot \rightarrow CH_3CH_2CH_2CH_3$$

$$CH_3\dot{C}HCH_3 + CH_3\cdot \rightarrow CH_3CH(CH_3)CH_3$$

$$CH_3\cdot + CH_3\cdot \rightarrow CH_3CH_3$$

$$CH_3CH_2CH_2\cdot + CH_3CH_2CH_2\cdot \rightarrow CH_3CH_2CH_2CH_2CH_2CH_3$$

$$CH_3\dot{C}HCH_3 + CH_3\dot{C}HCH_3 \rightarrow CH_3CH{=}CH_2 + CH_3CH_2CH_3$$

$$CH_3\dot{C}HCH_3 + CH_3\dot{C}HCH_3 \rightarrow CH_3CH(CH_3){-}CH(CH_3)CH_3$$

$$CH_3\dot{C}HCH_3 + CH_3CH_2CH_2\cdot \rightarrow CH_3CH(CH_3)CH_2CH_2CH_3$$

(The propyl radical pair and the isopropyl + propyl pair also give $CH_3CH{=}CH_2 + CH_3CH_2CH_3$, as indicated by arrows.)

Occurrence of the *direct* mechanism for insertion was established by Doering and Prinzbach, who studied the reaction of methylene with isobutene labelled at the $=CH_2$ group with ^{14}C (25). The 2-methylbutene-1 obtained in the gas-phase reaction had 92 per cent of its ^{14}C at the terminal unsaturated carbon atom, showing that 84 per cent of this compound was formed directly without rearrangement and only 16 per cent was formed by the radical mechanism.

$$^{14}CH_2{=}C(CH_3){-}CH_3 + CH_2 \xrightarrow{84\%} {}^{14}CH_2{=}C(CH_3){-}CH_2CH_3$$

$$\downarrow 16\%$$

$$\left[\begin{array}{c} ^{14}CH_2{=}C(CH_3){-}CH_2\cdot \\ \updownarrow \\ ^{14}\cdot CH_2{-}C(CH_3){=}CH_2 \end{array}\right] + CH_3\cdot \xrightarrow{8\%} CH_3{-}^{14}CH_2C(CH_3){=}CH_2 \quad (8\% \nearrow {}^{14}CH_2{=}C(CH_3){-}CH_2CH_3)$$

In the liquid phase even less rearrangement of the ^{14}C was observed.

Further evidence for the one-step nature of certain insertions is found in the observation that bicyclo[2.2.1]heptane is alkylated at the bridgehead as well as on the bridges (4).

$$C_7H_{12}\ (\text{bicyclo[2.2.1]heptane: } CH(CH_2CH_2)(CH_2)(CH_2CH_2)CH) + CH_2 \rightarrow CH_3{-}C(CH_2CH_2)(CH_2)(CH_2CH_2)CH$$

Franzen showed that liquid-phase methylation results in maintenance of configuration even in cases where the constraint of a bicyclic ring system is absent. In the case of both *cis*- and *trans*-cyclopentene diacetate the methylation that took place at the acetoxy-bearing carbon atoms gave no change in configuration.

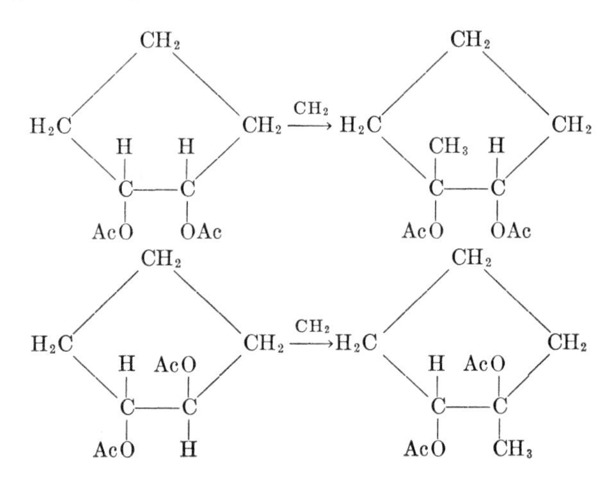

Reactivity in Methylene Insertions. The methylene molecules initially produced in the photolysis of diazomethane are extremely reactive species that possess far more than the thermal equilibrium energy characteristic of the temperature at which they are formed. Doering and coworkers showed that photolyzing diazomethane methylates pentane in the liquid phase almost randomly, not only near room temperature but even at $-75°$ (26).

$$CH_3CH_2CH_2CH_2CH_3 + CH_2 \xrightarrow{-75^\circ} \begin{array}{ll} 48\% & CH_3CH_2CH_2CH_2CH_2CH_3 \\ 35\% & CH_3CH(CH_3)CH_2CH_2CH_3 \\ 17\% & CH_3CH_2CH(CH_3)CH_2CH_3 \end{array}$$

Even more strikingly, in cyclohexene the vinyl, allyl, and "typical alicyclic" hydrogen atoms all show about the same reactivity. Methylene from the photolysis of ketene contains somewhat less excess energy and is therefore somewhat more selective in its reactions. Frey and Kistiakowsky found that the C_4H_{10} formed by the gas-phase photolysis of ketene in the presence of propane is 37 per cent isobutane instead of the 25 per cent that would have resulted from random methylation (23). They further showed that in the presence of an inert gas such as carbon dioxide or argon, to which the initially formed energy-rich methylene can give up (by collision) some of its excess energy, the selectivity increases, the yield of isobutane reaching 42 per cent in the presence of 64 mm. of carbon dioxide. Richardson, Simmons, and Dvoretzky pointed out that the preceding observation and others may be rationalized in terms of the assumption that insertion reactions in solution are ordinarily due to the non-selective one-step insertion by the initially formed singlet methylene, but that in the vapor phase there is a competing transformation of the singlet to the more selective triplet which gives insertion by the two-step mechanism via an intermediate radical (27). This assumption should be capable of being tested by a more complete and quantitative study than has yet been reported. In any event, it should be noted that, relative to the triplet, the singlet form of methylene has at least three advantages that should make it more facile at direct insertion. It is the more energetic form. Its direct insertions would not have to be accompanied by change in multiplicity. Its geometry (103° H—C—H bond angle) is much nearer that which it is required to assume in the reaction product (109.5° H—C—H bond angle).

Some interesting results of the excess energy carried by photochemically produced methylene may be seen in Frey's study of the methylation of cyclobutane (28). The methylcyclobutane initially produced is excited and contains enough excess energy to decompose readily to propylene and ethylene whenever enough of this excess energy becomes localized in the proper bonds. Before this happens, however, the excited molecule may lose so much of its excess energy by collision that it can no longer decompose. Thus the yield of methylcyclobutane increases as the pressure increases and is higher when the methylene is derived from ketene than it is when the more energetic methylene derived from diazomethane is used.

Cyclodiazomethane appears to yield methylene with less excess energy than ordinary linear diazomethane does under the same conditions. Frey and Stevens found that the photolysis of cyclodiazomethane using light of wave length 3130 Å in the presence of cyclobutane yields methylcyclobutane with essentially the same amount of excess energy as when linear diazomethane is photolyzed with light of wave length 3660 Å, which contains considerably less energy per quantum (29).

ADDITION OF METHYLENE TO OLEFINS

Doering and coworkers found in the reaction with cyclohexene that methylene adds to the double bond of an olefin in addition to giving insertion reactions (26). Using *cis*- and *trans*-2-butene, neither of which undergoes geometrical isomerization under the reaction conditions, Skell and Woodworth and Doering and LaFlamme showed that photolysis of diazomethane in the liquid phase or the gas phase at one-atmosphere pressure (with no added inert gas) results in addition of methylene to the double bond without change in the geometrical configuration of the methyl groups (30, 31).

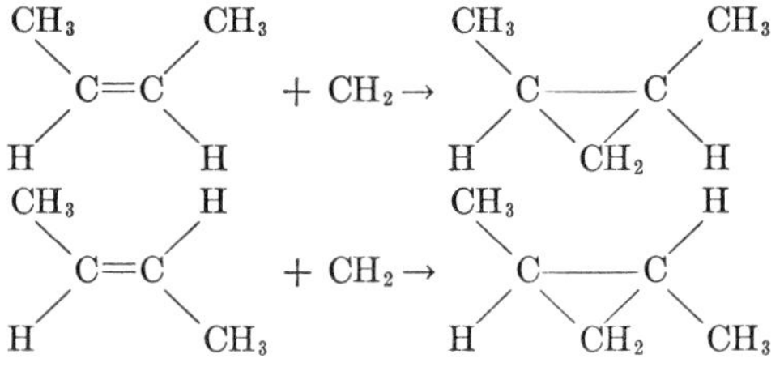

```
CH3        CH3             CH3              CH3
   \      /                   \            /
    C=C        + CH2 →         C ——————— C
   /      \                   /  \      /  \
H          H               H      CH2       H
CH3        H               CH3              H
   \      /                   \            /
    C=C        + CH2 →         C ——————— C
   /      \                   /  \      /  \
H          CH3             H      CH2       CH3
```

This result gave evidence for the singlet character of the reacting methylene since a triplet methylene would be expected to add like a radical to give an intermediate diradical that should lose its stereochemical homogeneity at a rate at least comparable to its rate of cyclization to dimethylcyclopropane.

```
CH3      CH3                  CH3      CH3
   \    /                        \ .   /
    C=C        + CH2 →            C—C
   /    \                        /  |\
  H      H                      H   | H
                                   ·CH2
                         ↙            ↓
 H       CH3               CH3              CH3
  \ .   /                     \            /
   C—C                         C——————————C
  /  |\                       /  \      /  \
CH3  | H                     H    CH2       H
    ·CH2
     ↓
H          CH3
 \        /
  C——————C
 /  \   / \
CH3  CH2   H
```

The overall stereochemical result of the addition of methylene to olefins and also the relative extents of addition and insertion are therefore at least somewhat dependent on whether the reacting methylene is in the singlet or triplet form. The multiplicity of the methylene is not the only relevant factor, however. There is also good evidence that in some cases the cyclopropanes initially formed by addition have so much excess energy that they may isomerize before they lose this excess energy by collision. Frey and Kistiakowsky found that methylene (from the photolysis of ketene) reacts with ethylene to give a mixture of cyclopropane and propylene with the yield of cyclopropane decreasing as the pressure in the mixture (consisting solely of ketene and ethylene) decreases (23). This observation may be explained either by the hypothesis that the initially formed "hot" cyclopropane isomerizes to propylene, or the hypothesis that the initially formed singlet methylene gives largely addition whereas the triplet methylene that is formed if the singlet does not collide with an ethylene molecule too soon gives largely insertion. The first hypothesis was shown to be the correct one by experiments in which an inert gas such as argon or carbon dioxide was added to the reaction mixture (in which ethylene was still present in considerable amounts). The addition of an inert gas should increase the rate of transformation of singlet to triplet methylene and also the rate of removal of excess energy from any "hot" cyclopropane; since the cyclopropane yield increased with the pressure of added gas it is deactivation of "hot" cyclopropane that is the important factor in this case. Similar results were obtained in the reaction of methylene with isobutene and *cis-* and *trans*-2-butene (32, 33), but in these cases the intermediate "hot" cyclopropanes, because of the larger number of internal vibrational modes among which the excess energy can be distributed, show much less tendency to isomerize. With *cis-* and *trans*-2-butene the intermediate hot cyclopropanes undergo both geometrical isomerism and (slower) structural isomerism.

$$(CH_3)HC{=}CH(CH_3) + CH_2 \rightarrow [\text{cyclo-}(CH_3)HC{-}CH_2{-}CH(CH_3)]^{*} \rightleftharpoons (CH_3)HC\cdot\ CH_2\ \cdot CH(CH_3)$$

$$(CH_3)HC\cdot\ CH_2\ \cdot CH(CH_3) \rightarrow CH_3CH{=}CHCH_2CH_3$$

$$(CH_3)HC\dot{C}{-}CH(CH_3)(\cdot CH_2) \rightarrow (CH_3)_2CH{-}CH{=}CH_2,\ CH_3CH_2C(CH_3){=}CH_2,\ CH_3CH{=}C(CH_3)_2, \text{ and } CH_3CH{=}CHCH_2CH_3$$

$$H(CH_3)\dot{C}\ CH_2\ \dot{C}H(CH_3) \rightarrow CH_3CH{=}CHCH_2CH_3$$

$$H(CH_3)C\dot{}{-}CH(CH_3)(\cdot CH_2) \rightleftharpoons [\text{cyclo-}H(CH_3)C{-}CH_2{-}CH(CH_3)]^{*}$$

The addition of triplet methylene to olefins can be observed if the concentration of the olefin is kept low enough (and particularly if an inert gas is present as a diluent) so that the initially formed singlet can change to a triplet before it reacts. Under these conditions the addition of methylene (from the photolysis of diazomethane) to the 2-butenes is no longer stereospecific; *trans*-2-butene gives considerable amounts of *cis*-1,2-dimethylcyclopropane and *cis*-2-butene may give mostly *trans*-1,2-dimethylcyclopropane (34, 35). In at least one of these studies the gas pressure was too high for any significant fraction of initially formed "hot" dimethylcyclopropane to have isomerized before deactivation (35).

Kopecky, Hammond, and Leermakers showed that methylene may be produced directly in the triplet form by a photosensitized decomposition (36). When benzophenone is the photosensitizing agent and light of wave length 3130 Å (at which diazomethane does not absorb appreciably) is used, the following sequence of reactions occurs:

$$(C_6H_5)_2CO \xrightarrow{h\nu} (C_6H_5)_2CO^* \text{ (singlet)}$$
$$(C_6H_5)_2CO^* \text{ (singlet)} \rightarrow (C_6H_5)_2CO^* \text{ (triplet)}$$
$$(C_6H_5)_2CO^* \text{ (triplet)} + CH_2N_2 \rightarrow (C_6H_5)_2CO + CH_2N_2^* \text{ (triplet)}$$
$$CH_2N_2^* \text{ (triplet)} \rightarrow CH_2 \text{ (triplet)} + N_2$$

Upon absorption of light the benzophenone is transformed initially to a short-lived excited singlet state which then rapidly decays to a longer-lived triplet. These triplets often undergo energy transfer with diazomethane molecules to give an excited triplet diazomethane that decomposes to nitrogen and triplet methylene. The presence of the more selective triplet methylene is shown by the fact that with cyclohexene the predominant reaction is addition to the double bond instead of insertion, and with *cis*- and *trans*-2-butene the addition reaction is no longer stereospecific.

Such metallic reagents as copper powder and ferric chelate compounds are known to catalyze the decomposition of diazomethane. The decomposition of diazomethane by the action of such catalysts in the presence of olefins has been found to lead only to the addition of methylene to the double bonds. This reaction is apparently not due to triplet methylene, however, since the olefin addition is stereospecific. The activity of the catalysts and the course of the reactions observed have been explained in terms of the formation of methylene complexes. Although little is really known about their structure the name "methylene-metal complex" helps remind one that their reactions bear considerable resemblance to those of methylenes.

Both Frey and Franzen studied the reaction of methylene with butadiene (37, 38). Among the reaction products are vinylcyclopropane and cyclopentene. Frey found that the cyclopentene yield increases considerably when the pressure is decreased showing that the initially formed energy-rich vinylcyclopropane, if not collisionally deactivated, may isomerize to cyclopentene. He also observed that cyclopentadiene is another product of the reaction and that its yield increases with decreasing pressure even more rapidly than the cyclopentene yield, which reaches a maximum at a fairly low pressure and then decreases when the pressure is lowered further. Apparently the cyclopentene also contains excess energy and decomposes to hydrogen and cyclopentadiene if not deactivated by collision.

$$CH_2 + CH_2{=}CH{-}CH{=}CH_2 \rightarrow CH_2{=}CH{-}\underset{\diagdown\ \ \diagup}{CH{-}\overset{*}{C}H_2}\atop CH_2$$

$$\rightleftarrows$$

$$\left[\begin{array}{c} CH_2{=}CH{-}\dot{C}H{-}CH_2{-}\cdot CH_2 \\ \updownarrow \\ CH_2{-}CH{=}CH{-}CH_2{-}\cdot CH_2 \end{array}\right] \quad \text{I}$$

$$\leftarrow \text{cyclopentene: } CH{=}CH,\ CH_2,\ \overset{*}{C}H_2,\ CH_2 \text{ (ring)}$$

$$\downarrow$$

$$H_2 + \text{cyclopentadiene: } CH{=}CH,\ CH,\ CH_2,\ CH \text{ (ring, } CH{=}CH \text{)}$$

It appears that even at infinite pressure the yield of cyclopentene would not drop to zero and therefore, as Franzen suggests, some of the cyclopentene is probably formed by the addition of triplet methylene to butadiene to give the resonance-stabilized diradical I directly.

THE REACTION OF METHYLENE IODIDE WITH ZINC IN THE PRESENCE OF OLEFINS

It is interesting that one of the more significant contributions to our knowledge of the nature of methylene intermediates comes from the study of a reaction that appears *not* to involve a methylene intermediate. Simmons and Smith found that when methylene iodide is treated with a zinc-copper couple in the presence of ether and an olefin a cyclopropane derivative is formed (39).

$$RCH{=}CHR + CH_2I_2 + Zn \rightarrow \underset{\diagdown\ \diagup \atop CH_2}{RCH\text{——}CHR} + ZnI_2$$

From this observation it would seem reasonable that the zinc acts to remove two iodine atoms from the methylene iodide leaving free methylene, which then adds to the olefin. Indeed, at one time, many investigators seemed to have assumed that, except for certain reactions in which a fairly stable carbanion may be written as an intermediate, e.g.,

$$ClCH_2CO_2Et \xrightarrow{EtO^-} Cl\overset{\ominus}{C}HCO_2Et$$

$$\downarrow CH_2{=}CHCO_2Et$$

$$\begin{array}{ccc} CH_2\text{——}CHCO_2Et & & CH_2\text{—}\overset{\ominus}{C}HCO_2Et \\ \diagdown\ \diagup & & | \\ CH & \leftarrow & ClCH \\ | & & | \\ CO_2Et & & CO_2Et \end{array}$$

the formation of a cyclopropane derivative from an olefin and another organic compound proved that the latter compound was intermediately transformed into a methylene. Simmons and Smith, however, found evidence that this is not so. Under their reaction conditions, cyclopropanes are formed stereospecifically with retention of the geometrical configuration of the olefin, and no products of insertion by methylene were observed. Yet if the active intermediate were triplet methylene the addition should not be stereospecific, and if it were singlet large amounts of insertion would be expected. Further evidence was obtained by adding the zinc-copper couple to methylene iodide in ether in the absence of olefins. It was thereby shown, in agreement with previous observations, that an organometallic compound is formed. The reaction solution can be filtered to obtain a copper-free solution of the organometallic compound which reacts with iodine to give methylene iodide and with water to give methyl iodide. Wittig and Schwarzenbach showed that the same organometallic compound, capable of adding methylene to olefins, can be prepared from diazomethane and zinc iodide (40). From these observations the compound must be either bis-(iodomethyl)-zinc or iodomethylzinc iodide.

$$CH_2I_2 + Zn \rightarrow (ICH_2)_2Zn \xrightarrow{I_2} CH_2I_2$$

$$(ICH_2)_2Zn \xrightarrow{H_2O} CH_3I$$

Simmons and Smith state that ether solutions of this organozinc compound are stable for hours but the addition of cyclohexene brings about the relatively rapid (and somewhat exothermic) formation of bicyclo[4.1.0]heptane.

$$(ICH_2)_2Zn + \text{cyclohexene} \rightarrow ZnI_2 + \text{bicyclo[4.1.0]heptane}$$

Since it seems implausible that the iodomethylzinc compound should decompose to methylene any faster in the presence of an olefin than in its absence, it was suggested that the cyclopropane is formed by a direct reaction of the olefin with the organometallic compound, perhaps involving a transition state like that shown in brackets as follows.

$$>C{=}C< \; + \; ZnI{-}CH_2{-}I \;\longrightarrow\; \left[\; C{\cdots}CH_2{\cdots}ZnI,\;\; C{\cdots}CH_2{\cdots}I \;\right] \;\longrightarrow\; >C{-}CH_2{-}C< \; + \; ZnI{-}I$$

Alternatively, in view of the known electrophilic character of X—Zn —Y compounds and the fact that the reactivity of olefins in the Simmons-Smith reaction is increased by electron-donating groups, it might be suggested that coordination of the pi electrons of the double bond with the zinc atom is of importance. Such coordination would increase the electrophilicity of both unsaturated carbon atoms of the olefin and increase their tendency to form a bond to the electron-rich carbon atom of the iodomethylzinc compound. Unfortunately, this entire problem is still somewhat unsettled, partly because the experimental support for Simmons and Smith's statement that an olefin accelerates the decomposition of the methylene iodide-zinc compound is not described in sufficient detail to make its validity easily assessable.

The methylene iodide-zinc compound has been referred to as "a complex of methylene with zinc iodide," and in this way the Simmons-Smith reaction can be spoken of as a methylene reaction. Considerable reference has been made to other complexes of methylene but in not all cases has it been clear whether the word "complex" is being used in a sense that would permit methane to be called a complex of methylene and hydrogen or not. In the present case, studies of the structure of the iodomethylzinc compound by various physical means might be useful, but above all a kinetic study of the cyclopropane-forming reaction is needed.

OTHER METHODS OF FORMING METHYLENE

The Reaction of Methyl Chloride with Phenylsodium. Friedman and Berger found that the reaction of methyl chloride with phenylsodium in a mixture of hexadecane and cyclohexene yields, after carbonation, 41 per cent toluene, 19 per cent benzene, 3 per cent bicyclo[4.1.0]heptane, 3 per cent benzoic acid, 2 per cent ethylbenzene, smaller amounts of *n*-propylbenzene, isopropylbenzene, and phenylacetic acid, and unreported amounts of ethylene and ethane (41). The replacement of the cyclohexene in the reaction mixture by other olefins led to the formation of the appropriate cyclopropane derivatives, with the addition being, in the case of *cis*-2-butene, stereospecifically *cis*. These observations were explained by the following reaction mechanism, in which methyl chloride undergoes α-dehydrochlorination to give methylene, which may add to the olefin or enter into a number of other reactions, as shown.

$$C_6H_5Na + CH_3Cl \begin{cases} \nearrow C_6H_6 + CH_2 + NaCl \\ \searrow C_6H_5CH_3 + NaCl \end{cases}$$

$$C_6H_5Na + CH_2 \rightarrow C_6H_5CH_2Na$$

$$C_6H_5CH_2Na + CH_3Cl \rightarrow C_6H_5CH_2CH_3 + NaCl$$

$$C_6H_5CH_2Na + CH_2 \rightarrow C_6H_5CH_2CH_2Na$$

$$C_6H_5CH_2CH_2Na + CH_3Cl \rightarrow C_6H_5CH_2CH_2CH_3 + NaCl$$

$$C_6H_5CH_2CH_2Na \rightarrow C_6H_5\underset{\substack{|\\ Na}}{C}HCH_3$$

$$C_6H_5\underset{\substack{|\\ Na}}{C}HCH_3 + CH_3Cl \rightarrow C_6H_5CH(CH_3)_2 + NaCl$$

Under the experimental conditions benzylsodium is said not to be formed by the reaction of toluene with phenylsodium.

Although the formation of the observed reaction products may thus be explained simply, there are certain objections to the hypothesis that methylene is an intermediate in this reaction. Triplet methylene should not have given stereospecific addition to *cis*-2-butene. Singlet methylene should give insertion reactions to an extent comparable to its addition to olefins, but the products of such insertion reactions were sought without success. The work of Simmons and Smith described in the previous section provides evidence that the formation of cyclopropane derivatives need not be considered proof for the formation of methylene intermediates and suggests that the tetra-covalent species $NaCH_2Cl$ may have been the species that reacted with the olefins. It might also be suggested that $NaCH_2Cl$ may react with organosodium compounds as follows,

$$C_6H_5Na + NaCH_2Cl \rightarrow C_6H_5CH_2Na + NaCl$$

but first, perhaps, the statement that phenylsodium does not react with toluene under the experimental conditions should be checked. Because of the colloidal character of many organosodium compounds in hydrocarbon solvents and the heterogeneous character of the reaction in question, one should add isotopically labelled toluene to the phenylsodium–methyl chloride reaction mixture and look for the isotopic label in the various reaction products.

The Reactions of Tetramethylammonium Salts with Strong Bases. Franzen and Wittig observed that tetramethylammonium bromide suspended in cyclohexene reacts with a one-to-ten mixture of phenyllithium and phenylsodium in ether to give 5–18 per cent bicyclo[4.1.0]heptane; phenyllithium alone is not effective (42). In view of the evidence that Wittig and coworkers have secured for the α-metallation of tetramethylammonium ions by phenyllithium, it seems quite probable that the first step of this reaction consists of α-metallation by sodium.

$$(CH_3)_4N^+ + Br^- + C_6H_5Na \rightarrow (CH_3)_3\overset{+}{N}CH_2Na + Br^- + C_6H_6$$

It has been suggested that this intermediate decomposes to trimethylamine (an observed by-product) and methylene, which then undergoes such subsequent reactions as addition to olefin molecules.

$$(CH_3)_3\overset{+}{N}CH_2Na \rightarrow Na^+ + (CH_3)_3\overset{+}{N}CH_2^- \rightarrow (CH_3)_3N + CH_2$$

If this is the case, decomposition to methylene must take place whether there is present any olefin with which the methylene can react or not. In the absence of an added olefin Wittig and Polster observed that tetramethylammonium bromide reacts with phenyllithium–phenylsodium (1:5) to give largely trimethylamine and a long-chain polymethylene polymer of m.p. 122–129° (43). For this polymethylene to have been formed by the polymerization of free methylene would require that the methylene show a very strong preference for reaction with the ends of the growing polymer chain rather than other possible reactants (such as the ether used as solvent) present in tremendously higher concentrations. In view of the fact that methylene does not usually display enormous selectivity in its reactions, it seems equally plausible to suggest that it is $(CH_3)_3\overset{+}{N}CH_2^-$ (or perhaps the related sodium compound) that adds methylene to cyclohexene, and perhaps also reacts with the ends of the growing polymethylene chain.

The question of whether reactions such as the preceding one actually involve the formation of free methylene as an intermediate need not concern the synthetic chemist who uses the reaction to prepare a new compound, of course, but it was the consideration of just such mechanistic questions that led to the creation of some highly useful synthetic procedures.

Miscellaneous Methods for Forming Methylene. By use of the tellurium-mirror method, Goldfinger, Le Goff, and Letort obtained evidence for the formation of methylene, perhaps in an electrically charged form, in the electric discharge-induced decomposition of ethylene, methylene bromide, and dioxane (44). Gevantman and Williams showed that a similar species may be formed by the radiolysis (by γ-rays) of methyl iodide (17). In the presence of radioactive iodine, methylene iodide and hydrogen iodide are reaction products.

$$CH_3I \rightarrow HI + CH_2 \xrightarrow{I_2^*} CH_2I_2^*$$

The almost complete absence of radioactivity from the hydrogen iodide formed rules out the possibility that the reaction is simply a substitution of iodine for a hydrogen atom.

Bawn and Tipper described evidence that methylene is formed as an intermediate in the vapor-phase reaction of sodium with methylene bromide (45).

$$Na + CH_2Br_2 \rightarrow NaBr + CH_2Br$$
$$Na + CH_2Br \rightarrow NaBr + CH_2$$

Knunyants, Gambaryan, and Rokhlin refer to a number of other suggestions that methylene is formed as an intermediate in various reactions (2). Most of these reactions were heterogeneous and it would be difficult to distinguish a free methylene intermediate from some CH_2-metal compound formed on the surface of the catalyst.

OTHER REACTIONS OF METHYLENE

Addition of Methylene to Itself and to Other Divalent Carbon Derivatives. The addition of such a simple species as methylene to itself might be expected to require a third body. That is, the ethylene molecule formed would obviously have the energy required to dissociate again (or to dissociate in some manner other than the reverse of that by which it was formed), and will do so unless it can give up some of its excess energy by collision with another molecule or the wall of the reaction vessel (or perhaps by radiation). Although rather large quantities of ethylene are often produced in reactions in which methylene is generated as an intermediate, in most cases this ethylene has been shown to arise from the attack of methylene on the initial reactant rather than from dimerization.

$$CH_2N_2 \rightarrow CH_2 + N_2$$
$$CH_2 + CH_2N_2 \rightarrow CH_2{=}CH_2 + N_2$$

The first evidence for the reaction of methylene with carbon monoxide to form ketene was obtained by Staudinger and Kupfer, who heated diazomethane in the presence of carbon monoxide (8). This reaction was studied in considerably more detail by Wilson and Kistiakowsky, who investigated the photolysis of ketene in the presence of ^{13}C-labelled carbon monoxide (46).

$$CH_2{=}CO \xrightarrow{h\nu} CH_2 + CO$$
$$CH_2 + {}^{13}CO \rightarrow CH_2{=}{}^{13}CO$$

The extent of formation of ^{13}C-labelled ketene increases with increasing pressure due to increased efficiency of deactivation of the initially formed hot $CH_2{=}{}^{13}CO$ molecules. By use of quantitative measurements it was shown that the ^{13}CO combines with methylene rather than activated ketene.

Reactions with Organic Halides. Urry and Eiszner discovered a remarkable reaction in which diazomethane, in the presence of light, transforms carbon tetrachloride into pentaerythrityl tetrachloride in 60 per cent yield (47). Similar results were obtained with other polyhalomethanes and with certain α-haloesters.

$$CCl_4 + 4CH_2N_2 \rightarrow C(CH_2Cl)_4 + 4N_2$$
$$BrCCl_3 + 4CH_2N_2 \rightarrow BrCH_2C(CH_2Cl)_3 + 4N_2$$
$$CHCl_3 + 4CH_2N_2 \rightarrow CH_3C(CH_2Cl)_3 + 4N_2$$
$$Cl_3CCOCH_3 + 3CH_2N_2 \rightarrow (ClCH_2)_3CCOCH_3 + 3N_2$$
$$BrCH_2CO_2CH_3 + CH_2N_2 \rightarrow BrCH_2CH_2CO_2CH_3 + N_2$$

The overall result of the reaction is to insert a methylene group between each halogen atom of the reactant and the carbon to which it had been attached. It was quickly shown, however, that the reaction does not involve consecutive one-step insertive attacks by free methylene. Such a reaction sequence [e.g., $CCl_4 \rightarrow Cl_3CCH_2Cl \rightarrow Cl_2C(CH_2Cl)_2 \rightarrow ClC(CH_2Cl)_3 \rightarrow C(CH_2Cl)_4$] was disproved by the observation that the possible intermediate, Cl_3CCH_2Cl, is inert under the reaction conditions. Instead, the reaction is believed to proceed by the following mechanism:

$$CH_2N_2 \xrightarrow{h\nu} CH_2 + N_2$$
$$CH_2 + CCl_4 \rightarrow \cdot CH_2Cl + Cl_3C\cdot$$
$$Cl_3C\cdot + CH_2N_2 \rightarrow Cl_3CCH_2\cdot + N_2$$
$$Cl_3CCH_2\cdot \rightarrow Cl_2\dot{C}CH_2Cl$$
$$Cl_2\dot{C}CH_2Cl + CH_2N_2 \rightarrow ClCH_2CCl_2CH_2\cdot + N_2$$
$$ClCH_2CCl_2CH_2\cdot \rightarrow (ClCH_2)_2\dot{C}Cl$$
$$(ClCH_2)_2\dot{C}Cl + CH_2N_2 \rightarrow (ClCH_2)_2CClCH_2\cdot + N_2$$
$$(ClCH_2)_2CClCH_2\cdot \rightarrow (ClCH_2)_3C\cdot$$
$$(ClCH_2)_3C\cdot + CH_2N_2 \rightarrow (ClCH_2)_3CCH_2\cdot + N_2$$
$$(ClCH_2)_3CCH_2\cdot + CCl_4 \rightarrow C(CH_2Cl)_4 + Cl_3C\cdot$$

The initiation steps (the first two steps shown) are plausible but are not as strongly demanded by the available evidence as is the rest of the reaction mechanism. The trichloromethyl radical, formed by this or some other method of initiation, is stabilized by its three α-chlorine substituents. It is able to react with diazomethane to give a simple primary and hence unstable radical because of the great driving force provided by the simultaneous formation of the very stable nitrogen molecule. This primary radical quite rapidly rearranges by migration of a β-chlorine atom to give a radical that is stabilized by two α-chlorine atoms and an α-chloromethyl substituent. This rearranged radical is stable enough to be selective and hence it reacts with diazomethane rather than taking part in a less thermodynamically favorable reaction with the far more abundant species carbon tetrachloride (the solvent). The reaction proceeds along these lines until the formation of the $(ClCH_2)_3CCH_2\cdot$, a primary radical that has no β-chlorine substituents and that therefore cannot facilely rearrange to a more stable radical. The highly reactive radical then reacts with the solvent to give the much more stable trichloromethyl radical and the entire cycle may then be repeated. In agreement with this proposed chain mechanism, Urry and Eiszner found the reaction to be inhibited strongly by diphenylamine.

Franzen (48) and Bradley and Ledwith (49) studied the photolysis of diazomethane in the presence of various alkyl halides. As principal products isopropyl chloride yields isobutyl chloride, and *t*-butyl chloride yields neopentyl chloride (48).

$$(CH_3)_2CHCl + CH_2N_2 \xrightarrow{h\nu} (CH_3)_2CHCH_2Cl$$

$$(CH_3)_3CCl + CH_2N_2 \xrightarrow{h\nu} (CH_3)_3CCH_2Cl$$

In view of the finding that *n*-propyl chloride is more reactive than isopropyl chloride which is more reactive than *t*-butyl chloride, the reactivity sequence: primary > secondary > tertiary, was reported for insertion reactions by methylene at the C—Cl bond (49). However, in view of the fact that we cannot be sure whether the reactions being studied were direct one-step insertions, two-step radical-coupling type insertions, reactions with a mechanism like that of the Urry-Eiszner reaction, or reactions proceeding via an ylide (cf. p. 139), it would be well to investigate these reactions further. It should be of interest to learn whether the reactions are subject to inhibition by free-radical inhibitors, whether reaction in the vapor phase gives the same products as that in the liquid phase, whether much of the *n*-butyl chloride formed from *n*-propyl chloride is formed by insertion into C—H bonds rather than into the C—Cl bond, and what products certain other possible reactants would yield.

Reactions of Methylene with Aromatic Compounds. As might be expected, the reaction of methylene with aromatic compounds may involve addition to the ring, insertion at a C—H bond, and perhaps other pathways as well. One of the products formed from the photolysis of diazomethane in benzene is toluene and the observation that the use of $^{14}CH_2N_2$ leads to $^{14}CH_3C_6H_5$, in which there is no ^{14}C in the aromatic ring, shows that there is no intermediate (such as certain cycloheptyl species) in which the entering carbon atom becomes equivalent to any of those that are already present (50).

A more interesting product of the photolysis of diazomethane in benzene is cycloheptatriene. This reaction was discovered by Doering and Knox (51); it is analogous to the formation of the Buchner acids by the decomposition of diazoacetic ester in benzene, which will be discussed in Chapter 7.

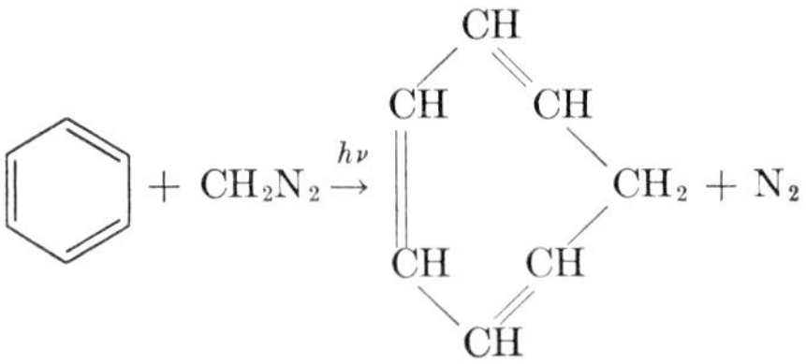

$$C_6H_6 + CH_2N_2 \xrightarrow{h\nu} C_7H_8 \text{ (cycloheptatriene)} + N_2$$

Evans and Lord summarized the evidence that there is interaction between the ends of the conjugated system, as Doering suggested, such that the contribution of a structure like the following one should be considered to explain the properties of cycloheptatriene (52).

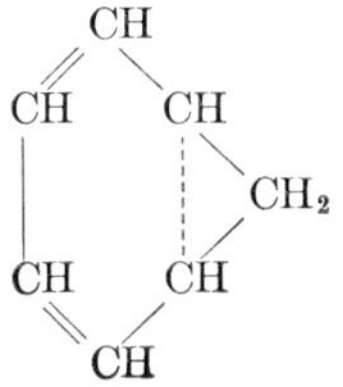

3

Dihalomethylenes

Dihalomethylenes were the first methylenes to be shown to be formed by heterolytic processes and, perhaps because of their ease of formation, have been investigated more thoroughly than any other type of substituted methylene. A claim of the preparation of dichloromethylene as a pure compound has been retracted, however.

FORMATION OF DIHALOMETHYLENES FROM HALOFORMS AND BASE

The Mechanism of the Basic Hydrolysis of Chloroform. There is good evidence that upon treatment with strong base chloroform undergoes the following reactions to give the reactive intermediate dichloromethylene, which then rapidly reacts further.

$$CHCl_3 + OH^- \rightleftharpoons CCl_3^- + H_2O$$
$$CCl_3^- \rightarrow CCl_2 + Cl^-$$

Evidence for the postulated initial carbanion formation may be found in the fact that chloroform undergoes certain base-catalyzed aldol condensations, e.g.,

$$CHCl_3 + OH^- \rightleftharpoons CCl_3^- + H_2O$$
$$\downarrow (CH_3)_2CO$$
$$CH_3\text{—}\underset{\underset{CH_3}{|}}{\overset{\overset{OH}{|}}{C}}\text{—}CCl_3 \xleftarrow{H_2O} CH_3\text{—}\underset{\underset{CH_3}{|}}{\overset{\overset{O^-}{|}}{C}}\text{—}CCl_3$$

and undergoes base-catalyzed deuterium exchange much more rapidly than it hydrolyzes.

One argument for dichloromethylene formation may be based on the fact that the relative reactivities of the chlorides of methane toward strong base cannot be rationalized in terms of the two generally accepted mechanisms (the S_N1 or carbonium ion mechanism and the S_N2 or direct bimolecular nucleophilic displacement mechanism) for halide hydrolysis (53). The basic decomposition of chloroform cannot be an S_N1 reaction since it is kinetically second-order, first-order in chloroform and first-order in hydroxide ions, and yet the reaction is much faster than would be plausible for the S_N2 mechanism. Methylene chloride is much less reactive toward nucleophilic reagents than is methyl chloride, showing that an α-chloro substituent decreases S_N2 reactivity. This decrease in reactivity toward nucleophilic reagents continues on going to chloroform if the nucleophilic reagents are sufficiently weakly basic. With strongly basic nucleophilic reagents, such as hydroxide and alkoxide ions, however, chloroform is far more reactive than methylene chloride and, toward hydroxide ions in aqueous solution, at least, even more reactive than methyl chloride. Such reactivity is readily explained in terms of the proposed reaction mechanism, in which the concentration of the intermediate trichloromethyl anion is proportional to the concentration and the strength of the base by whose action it is formed.

The strongest evidence for the intermediacy of dichloromethylene comes from kinetic studies in which capture of the intermediate by various nucleophilic reagents was demonstrated. It was found, for example, that although chloroform is almost inert to sodium thiophenoxide alone at 35°, in the presence of sodium hydroxide it reacts rapidly giving triphenylorthothioformate (53).

$$CHCl_3 \xrightarrow{OH^-} CCl_3^- \xrightarrow{-Cl^-} CCl_2$$

$$CCl_2 + C_6H_5S^- \rightarrow C_6H_5SCCl_2^- \xrightarrow[\text{steps}]{\text{several}} (C_6H_5S)_3CH$$

In other cases the entire CCl_2 group was maintained intact in the product of capture. Chloride ions were found to slow the basic hydrolysis of chloroform by capturing the intermediate dichloromethylene molecules, thus converting them to trichloromethyl anions and then back to chloroform (54). Bromide and iodide ions behave similarly but more effectively; they decrease the rate at which base is used up but in so doing bring about the transformation of much of the chloroform to bromodichloromethane and dichloroiodomethane, respectively, as shown in the following scheme:

$$CHCl_3 + OH^- \rightleftharpoons H_2O + CCl_3^-$$
$$CCl_3^- \rightleftharpoons CCl_2 + Cl^-$$
$$CCl_2 + Br^- \rightleftharpoons CCl_2Br^-$$
$$CCl_2Br^- + H_2O \rightleftharpoons CHCl_2Br + OH^-$$
$$CCl_2 + I^- \rightleftharpoons CCl_2I^-$$
$$CCl_2I^- + H_2O \rightleftharpoons CHCl_2I + OH^-$$
$$CCl_2 \xrightarrow[\text{several steps}]{OH^-,\ H_2O} CO + 2Cl^-$$

From the effect of known concentrations of the various halide ions on the rate of disappearance of base it was possible to calculate the relative rate constants for the combination of the halide ions with dichloromethylene. Comparison of these values with the *nucleophilicity constants* of Swain and Scott showed that the relative efficiencies with which the halide ions captured dichloromethylene varied in the same manner as the relative rates at which they performed other typical nucleophilic displacement reactions.

Horiuti, Tanabe, and coworkers suggested a somewhat different series of steps for the transformation of chloroform to dichloromethylene (55).

The Relative Ease of Formation of Dihalomethylenes from Haloforms. Although no other haloform has yet been studied in so much detail as has chloroform, a number of studies of the basic hydrolysis reactions give evidence, of the type already described for chloroform, for the intermediacy of dihalomethylenes in every case. As shown in Table 3–1 the reactivities, both in carbanion formation and for overall hydrolysis,

TABLE 3-1

Rate Constants for Base-Catalyzed Carbanion Formation and Hydrolysis by Haloforms (7)

	$10^5\ k$ in liters mole^{-1} sec^{-1} at 0° in water	
Haloform	Carbanion Formation	Hydrolysis
CHF_3	....	<0.00001
CHI_3	105,000	~0.001
$CHClI_2$	....	0.01
$CHCl_3$	820	0.06
$CHBrClI$	....	0.12
$CHBr_3$	101,000	0.24
$CHCl_2I$	4,800	0.26
$CHBr_2Cl$	25,000	0.66
$CHCl_2F$	16	1.23
$CHBrCl_2$	5,100	1.49
$CHClF_2$	[a]	1.7
$CHFI_2$	8,800	15
$CHBrClF$	365	132
$CHBrF_2$	[a]	208
$CHBr_2F$	3,600	277
CHF_2I	[a]	960

[a] These compounds give concerted α-dehydrohalogenation rather than carbanion formation.

vary over a wide range. The rates of carbanion formation may be rationalized fairly well by the generalization that the relative extent of stabilization of carbanions by α-halo substituents stands in the order $I \sim Br > Cl > F$. Rationalization of the relative hydrolysis rates is not so simple but it can be done satisfactorily in terms of the dihalomethylene reaction mechanism. The hydrolysis mechanism written earlier for the specific case of chloroform may be expressed more generally as follows:

$$CHXYZ + OH^- \underset{k_{-1}}{\overset{k_1}{\rightleftharpoons}} CXYZ^- + H_2O$$

$$CXYZ^- \xrightarrow{k_2} X\text{—}C\text{—}Y + Z^-$$

followed by rapid reaction of the X—C—Y, where X, Y, and Z are all halogen atoms that may or may not be different from each other. Application of the steady state treatment to the above reaction mechanism (in the case where recapture of X—C—Y by Z^- may be neglected) shows that the second-order rate constant obtained in the hydrolysis reaction (k_h) has the following significance in terms of the rate constants for the individual steps

$$k_h = \frac{k_1 k_2}{k_{-1} + k_2}$$

or, in the common case where $k_{-1} \gg k_2$,

$$k_h = \frac{k_1 k_2}{k_{-1}}$$

The relative values of k_1, the carbanion formation rate constant, have already been rationalized and it seems reasonable to assume that k_{-1}, the rate constant for carbanion protonation, will show a variation with structure that is, qualitatively, at least, just the opposite of that shown by k_1. Therefore all that remains is a rationalization of the relative values of the k_2's. In the formation of dihalomethylene the halogens are seen to have two different roles to play; one is to be lost as an anion (as Z in the mechanism above) and the other is to become a substituent (such as X and Y above) on the dihalomethylene. By the application of a semi-empirical linear free-energy relationship to the data on haloform hydrolysis, quantitative estimates were made of the relative abilities of the various halogens to play each of these roles (56). These estimates showed that the relative ease of loss as an anion (Br, I > Cl ≫ F) is about the same as that observed in a number of other organic reactions. The abilities of the various halogens to facilitate dihalomethylene formation, and hence presumably their abilities to stabilize dihalomethylenes, stand in the order: F ≫ Cl > Br > I. This order is believed to result in part from the relative abilities of the halogens to stabilize the electron-deficient carbon atom by use of their unshared electron pairs, as illustrated by the latter two of the three contributing structures below.

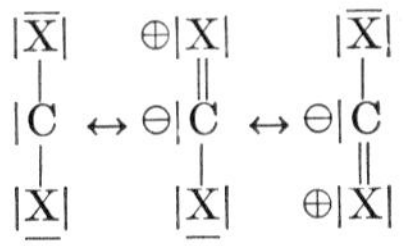

$$\begin{array}{ccccc} |\overline{X}| & & \oplus|X| & & |\overline{X}| \\ | & & \| & & | \\ |C & \leftrightarrow & \ominus|C & \leftrightarrow & \ominus|C \\ | & & | & & \| \\ |\underline{X}| & & |\underline{X}| & & \oplus|X| \end{array}$$

In view of the small bond angle (~103°) and hence the high degree of *p* character in the bonds of singlet methylene, it is probable that much of the stability of fluoromethylenes (relative to their precursors) is due to the decrease in the effective electronegativity of carbon that accompanies increases in the *p* character of the bonding; since the strength of the C—X bond increases with the square of the difference in electronegativity between C and X there is considerably more increase in bond energy with fluorine than with any other halogen. The difference in the stabilizing influence of two fluorine atoms and two iodine atoms corresponds to a 10^{10}-fold difference in rate of formation of dihalomethylenes at room temperature.

The Concerted Dehydrohalogenation of Haloforms. As described in the previous section fluorine is the poorest of all the halogens at stabilizing trihalomethyl anions and the best at stabilizing dihalomethylenes. These two factors work together to increase the probability that a trihalomethyl anion, once formed, will decompose to dihalomethylene rather than be reprotonated to haloform. Thus, for example, it may be seen from Table 3–1 that a tribromomethyl anion decomposes to dibromomethylene only about one time in every 421,000 that it is formed but that a dibromofluoromethyl anion decomposes to bromofluoromethylene one time in every thirteen. Extrapolation of these results might lead to the conclusion that the bromodifluoromethyl anion would yield difluoromethylene practically every time that it is formed so that the initial carbanion formation would be the rate-controlling step in the basic hydrolysis of bromodifluoromethane. Further examination of Table 3–1, however, reveals that the basic hydrolysis of bromodifluoromethane and also that of chlorodifluoromethane and difluoroiodomethane are all much faster than would seem reasonable for carbanion formation. Thus, in spite of the fact that the replacement of chlorine by fluorine decreases the rate of carbanion formation by from sevenfold (compare $CHBr_2Cl$ and $CHBr_2F$) to fifty-one-fold (compare $CHCl_3$ and $CHCl_2F$), with the magnitude of the effect increasing steadily with decreasing reactivity of the compounds involved, the rate of hydrolysis of bromodifluoromethane is more than half as large as the rate of carbanion formation by bromochlorofluoromethane. The most reasonable interpretation of these facts is that there has been a change in the reaction mechanism. With the three CHF_2X compounds in question the ease of dihalomethylene formation has become so great and trihalomethyl anion formation has become so difficult that the carbanion is by-passed and the dihalomethylene formed directly in a concerted α-dehydrohalogenation (57).

$$HO^- + H{-}\underset{F}{\overset{F}{\underset{|}{\overset{|}{C}}}}{-}Br \rightarrow \overset{-\delta}{HO}\cdots H\cdots\underset{F}{\overset{F}{\underset{|}{\overset{|}{C}}}}\cdots\overset{-\delta}{Br} \rightarrow HOH + \underset{F}{\overset{F}{\underset{|}{\overset{|}{C}}}} + Br^-$$

In agreement with this mechanism the basic hydrolysis of deuteriobromodifluoromethane has been found to be unaccompanied by deuterium exchange with the solvent.

ADDITION OF DIHALOMETHYLENES TO OLEFINS

One of the most useful reactions of dihalomethylenes is addition to olefins, discovered by Doering and Hoffmann. These workers found that the addition of chloroform, bromoform, or iodoform to a cold suspension of potassium *t*-butoxide in cyclohexene and certain other olefins brought about the formation of a 1,1-dihalocyclopropane derivative (58). Experiments with *cis* and *trans* olefins showed that the addition is cleanly stereospecific, with retention of geometrical configuration (31, 59). This stereospecificity shows that the reaction very probably does not involve the intermediate formation of a zwitterion or diradical but is probably a one-step process in which both new bonds are being formed simultaneously (but may not be formed to the same extent in the transition state).

CH_3 CH_3
C=C $+ CCl_2$ $\not\rightarrow$ CH_3 $\overset{\oplus}{C}$—C CH_3
H H H H
$\ominus CCl_2$

CH_3 $\dot{C}$—C CH_3 CH_3 C——C CH_3
H H H CCl_2 H
$\cdot CCl_2$

In view of the evidence obtained in the case of methylene that the singlet undergoes one-step addition and the triplet adds via an intermediate diradical, it seems likely that the dihalomethylenes are singlets and that their additions to olefins do not proceed via intermediate diradicals. Nevertheless additional evidence against the diradical mechanism as well as the zwitterion mechanism has been obtained by competition experiments in which a dihalomethylene is generated in the presence of a mixture of two olefins, and the reaction products are analyzed to determine the extent to which each of the available olefins has reacted. Some of the available data, due to Skell and Garner (60) and Doering and Henderson (61), are listed in Table 3–2. It is seen that the relative reactivities of the olefins toward dibromomethylene and dichloromethylene are not at all like the reactivities toward the trichloromethyl radical, which adds to give a radical, nor the reactivity toward the hydrogen ion, which adds to give a carbonium ion, but there is a rough parallel with the reactivities toward bromine, an electrophilic reagent that forms two new bonds simultaneously to the two unsaturated carbon atoms of the olefins giving a three-membered ring (a bromonium ion). The behavior of dihalomethylenes as electrophilic reagents in the present case is analogous to their behavior toward halide ions and water.

TABLE 3–2

Relative Reactivities of Olefins Toward Various Species

Olefin	$k_{olefin}/k_{isobutylene}$				
	CCl_2	CBr_2	$\cdot CCl_3$	Br_2	H^+
$(CH_3)_2C{=}C(CH_3)_2$	6.6	3.5		2.5	
$(CH_3)_2C{=}CHCH_3$	2.9	3.2	0.17	1.9	0.58
$(CH_3)_2C{=}CH_2$	1.00	1.00	1.00	1.00	1.00
$CH_3CH{=}CHC_2H_5$ (*trans*)	0.26				<0.001
$CH_2{=}CHOC_2H_5$	0.23				9000
$CH_2{=}CH{-}CH{=}CH_2$		0.5	>40		
Cyclohexene	0.12	0.4	0.045		
$C_6H_5CH{=}CH_2$		0.4	>20	0.59	
n-$C_4H_9CH{=}CH_2$	0.023	0.07	0.19	0.36	

THE STRUCTURE OF DIHALOMETHYLENES

Venkateswarlu observed that an uncondensed discharge (5,000–10,000 volts) in carbon tetrafluoride at about 0.1 mm. pressure gives rise to an emission spectrum containing a band system in the region 2400–3250 Å that can reasonably be attributed only to a non-linear triatomic molecule, presumably difluoromethylene (62). Laird, Andrews, and Barrow used discharge through a "fluorocarbon" vapor (63) and Mann and Thrush used flash photolysis (64) in obtaining the absorption spectrum of this species. All three groups of workers agree that the vibrational frequencies deduced from the spectra are reasonable for difluoromethylene, and the latter two groups pointed out that the non-linear difluoromethylene observed must be the ground state. It therefore appears that for difluoromethylene, unlike methylene itself, the most stable state is a singlet rather than a triplet. This is to be expected in view of the fact that all other non-hydride triatomic species containing 18 outer-shell electrons have non-linear singlet ground states. The analogy is particularly relevant for species like O_3, NO_2^-, and NOF, in which all three atoms have one and only one inner shell of electrons.

There is reason to believe that the stability of the singlet form relative to the triplet should be greater for difluoromethylene than for any of the other dihalomethylenes. Since bond energies are known to increase with the square of the difference in electronegativities of the bonded atoms there will be a tendency for a carbon atom bound to fluorine to assume the valence state in which it has the least electronegativity; that is, of various possible *s-p* hybrid states, the states containing more *p* and less *s* character will be preferred. Therefore the linear triplet state, with *sp* hybridization (50 per cent *s* character) will not be so favorable (as far as the electronegativity of the carbon is concerned) as a non-linear singlet state in which carbon's orbitals are much more highly *p* in character. This tendency, which, for an electronegativity difference of 0.1 for two states of carbon, would amount to about 6.7 kcal./mole for difluoromethylene, diminishes rapidly in the series dichloromethylene, dibromomethylene, diiodomethylene.

As described previously the singlet forms of dihalomethylenes should be stabilized by the contribution of such structures as

$$|\overset{\oplus}{\overline{X}} = \overset{\ominus}{\overline{C}} - \underline{\overline{X}}|$$

with the amount of such stabilization decreasing in the order $F > Cl > Br > I$. Furthermore, in view of the fact that α-halogen substituents appear to stabilize free radicals in the order $I > Br > Cl > F$ we should expect iodine to stabilize the triplet form most and fluorine to stabilize it least (56).

From these facts it seems possible that the multiplicity of the ground state might change at some point in the series CF_2, CCl_2, CBr_2, CI_2. The change does not appear to have occurred with either dichloromethylene or dibromomethylene. As stated in the preceding section both these species add stereospecifically to olefins. Furthermore this stereospecific addition certainly does not appear to be a rapid diffusion-controlled reaction of an unstable initially formed singlet that has not yet decayed to a more stable triplet. Because of the presence of the fairly heavy chlorine and bromine nuclei, changes in multiplicity of these dihalomethylenes could be relatively rapid, and yet the substantial amount of selectivity found in reactions with olefins shows that these dihalomethylenes usually undergo a number of collisions before reaction. A possible loophole in this argument is the fact that the heavy nuclei present may make multiplicity changes so rapid that the diradical formed by addition of triplet dihalomethylene to an olefin cyclizes faster than it undergoes rotation around its carbon-carbon single bonds. The relative reactivities of olefins, however, are not what would be expected for addition by a diradical.

In the case of diiodomethylene, the dihalomethylene most likely to have a triplet ground state, there appears to be very little relevant evidence. Diiodomethylene has been reported to be formed in the basic decomposition of iodoform (56) and it appears to add to cyclohexene (58) but the adduct has not been fully characterized. The stereochemistry of its addition to olefins does not seem to have been studied, however.

From the temperature at which the spectrum of difluoromethylene becomes observable in carbon tetrafluoride-graphite mixtures the heat of formation of difluoromethylene has been estimated to be -35 ± 10 kcal./mole (65). This corresponds to a heat of atomization of about 242 kcal./mole and hence an average C—F bond energy of 121 kcal./mole, which is greater than the 115 kcal. average C—F bond energy of carbon tetrafluoride.

OTHER METHODS OF FORMING DIHALOMETHYLENES

Inasmuch as the transformation of haloforms to trihalomethyl anions may lead to the formation of dihalomethylenes it is not surprising that a number of other reactions in which trihalomethyl anions would be expected to be generated also yield dihalomethylenes. A number of such methods of generating dihalomethylenes, as well as a few methods that are not based on this principle, have been developed.

Dihalomethylenes from Trihaloacetate Anions. It has been long recognized that the decarboxylation reactions of trichloroacetic acid and tribromoacetic acid are first-order reactions of the anions leading initially to carbon dioxide and a trihalomethyl anion, which, in the hydroxylic solvents ordinarily employed, then rapidly abstracts a proton from the solvent to yield a molecule of haloform.

$$Cl_3CCO_2^- \rightarrow CO_2 + CCl_3^- \xrightarrow{ROH} CHCl_3$$

Wagner heated sodium trichloroacetate in a mixture of ethylene glycol dimethyl ether (to increase the solubility of the salt) and cyclohexene and obtained 7,7-dichlorobicyclo-[4.1.0]heptane in 65 per cent yield, a considerably higher yield than when the intermediate dichloromethylene is generated from chloroform and potassium *t*-butoxide (66, 67). Under these conditions, no doubt, the trichloroacetate anions, as usual, yield trichloromethyl anions, but *these* anions, having no fairly acidic protons with which to combine, can only decompose to dichloromethylene. The dichloromethylene molecules combine rather efficiently with cyclohexene, since the weakly basic trichloroacetate ions are not very nucleophilic. It is probable that when dichloromethylene is generated via potassium *t*-butoxide many of the dichloromethylene molecules are captured by the strongly basic alkoxide ions. Judging from the even lower yields of olefin-addition products obtained when sodium and potassium methoxide and ethoxide are used, steric factors may also influence the relative reactivity of various nucleophilic reagents toward dihalomethylenes.

The similarity between the loss of a proton by a haloform and the loss of carbon dioxide by a trihaloacetate anion has also been used in securing additional evidence for the concerted formation of difluoromethylene. Earlier in this chapter is described evidence that when a base removes a proton from bromodifluoromethane, chlorodifluoromethane, or difluoroiodomethane, a halide anion is lost and difluoromethylene is formed in one step without the intervention of a trihalomethyl anion. By analogy it would be expected that the loss of carbon dioxide from an $XCF_2CO_2^-$ anion (where X is Cl, Br, or I) would lead directly to difluoromethylene without the formation of an intermediate carbanion.

$$X{-}CF_2{-}CO_2^- \rightarrow \left[\overset{-\delta}{X}\cdots\overset{-\delta}{C}F_2\cdots\overset{-\delta}{C}O_2\right] \rightarrow X^- + CF_2 + CO_2$$

transition state

Therefore it would be expected that, unlike all the trihaloacetic acids that had been studied previously, chloro- (or bromo- or iodo-) difluoroacetic acid would *not* yield the corresponding haloform on decarboxylation (unless some is formed from the difluoromethylene). The experimental facts are in agreement with the requirements of the concerted mechanism. The decarboxylation of 0.13 *M* chlorodifluoroacetate leads to the formation of only four per cent haloform, and three-fourths of this haloform is fluoroform (68). This fluoroform arises from the combination of difluoromethylene with a fluoride ion (and a proton), as is shown by the sharp increase in the fluoroform yield when sodium fluoride is added to the reaction solution. It could not have arisen from any trifluoroacetic acid present as an impurity in the reactant because trifluoroacetic acid does not decarboxylate appreciably under the reaction conditions. Similarly the one per cent of chlorodifluoromethane formed must arise, largely, at least, from the combination of difluoromethylene with chloride ions (and a proton) because the yield increases greatly when sodium chloride is added to the reaction mixture.

$$ClCF_2CO_2^- \rightarrow Cl^- + CF_2 + CO_2$$
$$CF_2 + 2H_2O \rightarrow HCO_2H + 2HF$$
$$CF_2 + H_2O \rightarrow CO + 2HF$$
$$CF_2 + F^- \rightarrow CF_3^- \rightarrow CHF_3$$
$$CF_2 + Cl^- + H^+ \rightarrow CHClF_2$$

It should be noted that this combination of difluoromethylene with a chloride ion and a proton (perhaps from H_2O or perhaps from H_3O^+) must be a concerted process, not involving an intermediate trihalomethyl anion. This follows from the principle of microscopic reversibility. If the transformation of chlorodifluoromethane to difluoromethylene is a concerted reaction under a given set of conditions, the reverse reaction must also be concerted. The transformation of difluoromethylene to fluoroform, in contrast, is written as a two-step reaction involving an intermediate trifluoromethyl anion. The evidence for this is the fact that fluoroform may be produced in good yield in the decarboxylation of trifluoroacetic acid, and it is therefore likely that this decarboxylation proceeds via an intermediate trifluoromethyl anion. If the removal of carbon dioxide from a trifluoroacetate ion yields a trifluoromethyl anion then the removal of a proton from fluoroform will probably do so, too, and if the transformation of fluoroform to difluoromethylene passes through an intermediate carbanion so does the reverse reaction.

The decarboxylation experiments provide a somewhat better test of the concerted reaction mechanism than does the observation that the basic hydrolysis of deuteriobromodifluoromethane is not accompanied by deuterium exchange. Some of the decarboxylation experiments were carried out at hydrogen-ion concentrations more than 10^{10} times as large as those used in the hydrolysis experiments and under these more acidic conditions an intermediate carbanion would be more likely to be protonated. This statement is based on the assumption that in acidic solutions, at least, the more acidic hydrogen ions should compete effectively, as protonating agents, with the more abundant solvent. This is known to be the case with the trichloromethyl anion. Kinetic measurements show that at pH 4 the deuterium exchange of chloroform is catalyzed to about the same extent by water as by hydroxide ions (69). From this observation and the principle of microscopic reversibility it follows that at pH 4 trichloromethyl anions are protonated to about the same extent by water and by hydrogen ions, and therefore at pH 1 they are protonated about 1,000 times as fast as they would be in a basic solution.

Further evidence for the concerted mechanism for the formation of difluoromethylene in the decomposition of chlorodifluoroacetic acid may be found in the relative rates of decomposition of the four $Cl_nF_{3-n}CCO_2^-$ anions. The values of log k (in sec^{-1}) in water at 70° are −4.8, −6.8, −7.1, and −10.6 for $Cl_3CCO_2^-$, $Cl_2CFCO_2^-$, $ClCF_2CO_2^-$, and $CF_3CO_2^-$, respectively. These data would fit a simple linear free-energy relationship, in which every replacement of chlorine by fluorine decreases log k by about 1.9, except that $ClCF_2CO_2^-$ is about forty times as reactive as would be required. This anomalously high reactivity suggests that, unlike the other three anions, the chlorodifluoroacetate anions do not decompose to trihalomethyl anions. Instead they have available to them a more favorable reaction path, the concerted formation of difluoromethylene.

The decarboxylation of sodium chlorodifluoroacetate in diethylene glycol dimethyl ether has been used as a method for adding difluoromethylene to olefins (70).

Dihalomethylenes from Trihaloacetate Esters and Trihalomethyl Ketones. Parham and coworkers found that esters of trichloroacetic acid react with sodium or potassium alkoxides in the presence of olefins to give 1,1-dihalocyclopropanes in yields (80 ± 8% in the cases studied) that are higher than those obtained when chloroform is used as the source of dichloromethylene (71). The reaction is initiated by the attack of alkoxide ion on the carbonyl group to give an intermediate anion that may revert to product or cleave to dialkyl carbonate and trichloromethyl anion.

$$Cl_3CCO_2R + RO^- \rightleftharpoons Cl_3C\text{—}C(O^\ominus)(OR)_2$$

$$\downarrow$$

$$R_2C\text{——}CH_2 \text{ (cyclopropane ring with } CCl_2) \xleftarrow{R_2C=CH_2} CCl_2 \xleftarrow{-Cl^-} Cl_3C^- + (RO)_2CO$$

The lower yield obtained by use of chloroform and alkoxide ions may be due partly to the reaction of some of the intermediate dichloromethylene with the alcohol formed as a by-product.

Potassium *t*-butoxide and *t*-butyl dichloroacetate react with olefins to give 1,1-dichlorocyclopropanes in rather poor yield (71, 72). Parham and coworkers described evidence that this reaction involves the disproportionation of the starting ester to *t*-butyl *tri*chloroacetate.

$$Cl_2CHCO_2Bu\text{-}t + t\text{-}BuO^- \rightleftharpoons Cl_2\overset{\ominus}{C}CO_2Bu\text{-}t + t\text{-}BuOH$$

$$Cl_2CHCO_2Bu\text{-}t + Cl_2\overset{\ominus}{C}CO_2Bu\text{-}t \rightarrow Cl\overset{\ominus}{C}HCO_2Bu\text{-}t + Cl_3CCO_2Bu\text{-}t$$

Hexachloroacetone and alkali-metal alkoxides have also been used as sources of dichloromethylene (73, 74).

Dihalomethylenes from Tetrahalomethanes and Organometallic Compounds. Tetrahalomethanes rather commonly undergo nucleophilic attack on halogen with the displacement of a trichloromethyl anion. Miller and Kim used a reaction of this sort, the reaction of tetrahalomethanes with alkyl lithium compounds, to generate dihalomethylenes, which were then captured by olefins (75). With butyllithium, bromotrichloromethane, and cyclohexene in ether at −30°, 7,7-dichlorobicyclo[4.1.0]heptane was obtained in 91 per cent yield.

$$n\text{-}BuLi + BrCCl_3 \rightarrow n\text{-}BuBr + LiCCl_3$$
$$LiCCl_3 \rightarrow LiCl + CCl_2$$
$$CCl_2 + C_6H_{10} \rightarrow C_7H_{10}Cl_2$$

As implied in the reaction scheme above, the conjugate cation probably plays an important role in the decomposition of a trihalomethyl anion in such a poor ion-solvating medium as ether. In such a solvent, in fact, the line between those dihalomethylene-forming reactions that pass through an intermediate carbanion and those that are concerted processes may well be different from the one to be drawn between these two types of reaction mechanisms in aqueous solution.

Franzen has used kinetic measurements on a reaction of this type to secure strong additional evidence for the formation of free difluoromethylene as a reaction intermediate (76). He pumped separate streams of ethereal *n*-butyllithium and ethereal bromotrifluoromethane into a mixing nozzle and allowed the resultant stream to flow rapidly through a capillary into a larger tube through which cyclohexene was flowing. The reaction product was 7,7-difluorobicyclo[4.1.0]heptane. When a methanol-ether mixture rather than cyclohexene was pumped through the large tube, no significant amount of lithium methoxide was formed. This shows that the reaction of butyllithium with bromotrifluoromethane was complete even during the short time of flow (as little as 10^{-4} second in some cases) through the capillary. If the plausible assumption that trifluoromethyllithium would also react with methanol to give lithium methoxide (and fluoroform, which is relatively unreactive toward alkali) is true, it also follows that no trifluoromethyllithium remains. Hence the difluorobicycloheptane must have been formed from free difluoromethylene. By varying the flow rate, the half-life of difluoromethylene was found to be between 5×10^{-4} and 10^{-3} second under the conditions employed. With too slow a flow rate the difluoromethylene dimerized to tetrafluoroethylene before it came in contact with the olefin.

Hauser and coworkers have generated dichloromethylene by the action of benzhydrylpotassium on carbon tetrachloride and bromotrichloromethane as well as chloroform (77).

Dichloromethylene in the Pyrolysis of Chloroform. Semeluk and Bernstein showed that the pyrolysis of chloroform at 450–525° yields largely hydrogen chloride and tetrachloroethylene, and that the reaction is homogeneous, first-order in chloroform (at small extents of reaction, at least), and is slowed by the hydrogen chloride produced (78). They also found the reaction to be subject to a significant deuterium kinetic isotope effect, and the pyrolysis of $CHCl_3$ in the presence of DCl was found to give considerable $CDCl_3$.

Shilov and Sabirova verified some of these observations and also found that the reaction is not significantly inhibited by toluene (79). In view of the lack of inhibition by toluene it seems unlikely that the decomposition is a free-radical chain reaction, and in view of the deuterium kinetic isotope effect (k_H/k_D is 1.65 at 574°) it seems quite unlikely that the rate-controlling step is a simple cleavage of a carbon-chlorine bond.

$$CHCl_3 \rightarrow \cdot CHCl_2 + Cl\cdot$$

Another argument, based on the relative reactivities of other chlorides of methane, has been advanced against the free-radical rate-controlling step for the decomposition of chloroform shown above. Shilov and Sabirova presented evidence that the pyrolysis of methyl chloride and the pyrolysis of carbon tetrachloride in the presence of toluene both involve, as the rate-controlling step, the radical cleavage of a carbon-chlorine bond (80).

$$CCl_4 \rightarrow \cdot CCl_3 + Cl\cdot$$
$$CH_3Cl \rightarrow \cdot CH_3 + Cl\cdot$$

The decomposition of carbon tetrachloride is much faster than that of methyl chloride, as would be expected from the large amount of evidence that α-chlorine substituents (relative to α-hydrogens) stabilize free radicals. This difference in reactivity amounts to about 64,000-fold at 600° (based on the activation energies and frequency factors reported). Assuming that each new α-chlorine atom increases the reactivity by about twenty-five times, the radical cleavage of a carbon-chlorine bond in chloroform should be only about $1/25$ as fast as the same process in carbon tetrachloride. Experimentally, however, it is found that chloroform decomposes about twenty times as fast as carbon tetrachloride and thus more than 600 times as fast as would be expected from the free-radical mechanism. It therefore appears that for chloroform there is another reaction path more favorable than radical cleavage of a carbon-chlorine bond. Shilov and Sabirova have pointed out that this path is quite probably a cleavage to hydrogen chloride and dichloromethylene.

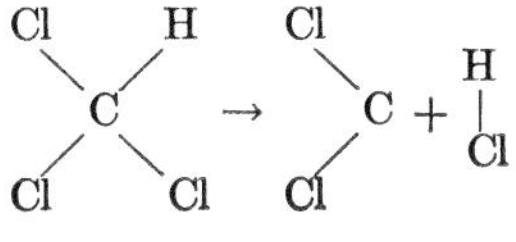

$$CHCl_3 \rightarrow CCl_2 + HCl$$

Although our knowledge of the nature of the subsequent reactions of dichloromethylene is not at all complete, recombination with hydrogen chloride must be common. This explains the inhibiting effect of added hydrogen chloride and the deuterium exchange with added deuterium chloride.

In view of the evidence that the homogeneous gas-phase first-order thermal β-dehydrohalogenation of a number of organic halides involves the intermediate formation of a carbonium-halide ion-pair, it is interesting to wonder whether the pyrolysis of chloroform does also.

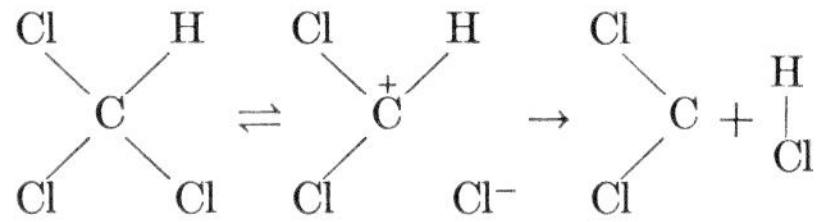

$$Cl_2CHCl \rightleftharpoons Cl_2CH^{+}\ Cl^{-} \rightarrow Cl_2C + HCl$$

Dihalomethylenes from Miscellaneous α-Elimination Reactions. Dihalomethylenes may be generated from carbanions other than trihalomethyl anions. Difluoromethyl phenyl sulfone reacts with base to give an intermediate carbanion, which loses a benzenesulfinate anion to give difluoromethylene (81). When relatively electropositive atoms are attached to trihalomethyl groups there is a strong tendency for α-elimination. Seyferth and coworkers took advantage of this fact in devising a method for the generation of dihalomethylenes from phenyltrihalomethylmercury compounds (82). Because of the relatively weakly nucleophilic character of the one reagent used, this method of generating dihalomethylenes appears to provide the best method currently available for achieving the reaction of these intermediates with very weakly nucleophilic reactants. Cyclopropane derivatives were obtained in greater than 65 per cent yield from ethylene, stilbene, and tetrachloroethylene, none of which had been found to react with haloforms in the presence of potassium *t*-butoxide.

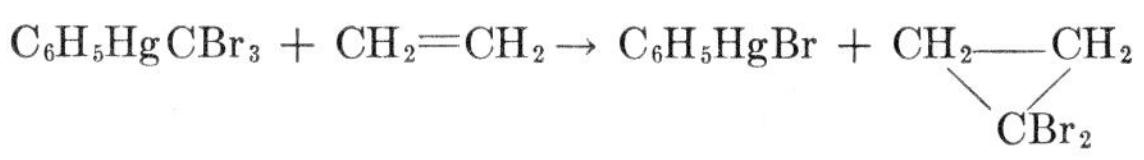

$$C_6H_5HgCBr_3 + CH_2{=}CH_2 \rightarrow C_6H_5HgBr + \underset{\diagdown\ \diagup}{CH_2\text{—}CH_2} \atop CBr_2$$

Nenitzescu and coworkers secured evidence (addition to cyclohexene) for the formation of dihalomethylenes in the reaction of silver nitrate with chloroform and bromoform and in the decomposition of silver trichloroacetate (83).

In these cases the departing halide ion may be not only pushed off carbon by the partial negative charge on the carbon but also pulled off by the mercury or silver ion.

Mahler has described evidence that $(CF_3)_3PF_2$ decomposes to yield difluoromethylene at temperatures around 100°.

OTHER REACTIONS OF DIHALOMETHYLENES

Addition of Dihalomethylenes to Various Multiple Bonds. Dihalomethylenes have been found to add to the multiple bonds of several types of compounds in addition to the simple olefins discussed earlier in this chapter. In at least one case, that of 2-methylpentene-1-yne-3, of a molecule with both a double and a triple bond, the only product isolated resulted from addition of dichloromethylene to the double bond, the preferred point of attack of most electrophilic reagents (84).

$$CH_3C{\equiv}C{-}\underset{\displaystyle CH_3}{\underset{|}{C}}{=}CH_2 \xrightarrow{CCl_2} CH_3C{\equiv}C{-}\underset{\displaystyle CH_3}{\underset{|}{C}}\overset{CCl_2}{\diagup\!\diagdown}CH_2$$

With 2-methylhexene-2-yne-4, however, addition appears to have occurred to a significant extent at each of the multiple bonds (84).

Addition of dihalomethylenes to acetylenes is of particular interest in that it provides the best method of synthesis of cyclopropenones. Kursanov, Volpin, and Koreshkov found that the reaction of diphenylacetylene with bromoform and potassium *t*-butoxide followed by addition of water gives diphenylcyclopropenone in 28 per cent yield, presumably via the intermediate formation of diphenyldibromocyclopropene (85).

$$C_6H_5{-}C{\equiv}C{-}C_6H_5 + CBr_2 \rightarrow \left[C_6H_5{-}\underset{\diagdown CBr_2 \diagup}{C{=\!=}C}{-}C_6H_5 \right]$$

$$\downarrow t\text{-BuO}^-$$

$$C_6H_5{-}\underset{\diagdown CO \diagup}{C{=\!=}C}{-}C_6H_5 \xleftarrow{H_2O} \left[C_6H_5{-}\underset{\diagdown t\text{-BuO}{-}C{-}OBu\text{-}t \diagup}{C{=\!=}C}{-}C_6H_5 \right]$$

Breslow and Peterson prepared di-*n*-propylcyclopropenone analogously from di-*n*-propylacetylene (86).

Mahler has added two consecutive molecules of difluoromethylene to the triple bond in hexafluoro-2-butyne.

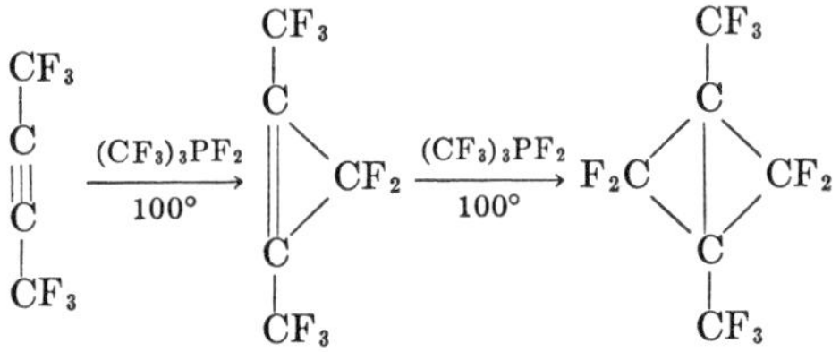

$$CF_3-C\equiv C-CF_3 \xrightarrow[100^\circ]{(CF_3)_3PF_2} \text{(cyclopropene: } CF_3C{=}CCF_3 \text{ bridged by } CF_2) \xrightarrow[100^\circ]{(CF_3)_3PF_2} \text{(bicyclobutane: } CF_3C{-}CCF_3 \text{ bridged by two } CF_2 \text{ groups, } F_2C \text{ and } CF_2)$$

The intermediate fluorinated cyclopropene was isolated.

As further examples of the wide variety of olefins to which dichloro- or dibromomethylene has been added, several ketene acetals (87) and allenes (88) may be mentioned.

Dichloromethylene has also been reported to add to the carbon-nitrogen double bond of benzalaniline (89).

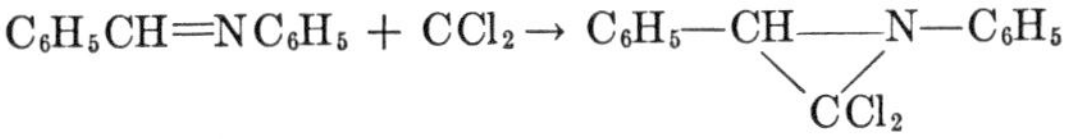

$$C_6H_5CH{=}NC_6H_5 + CCl_2 \rightarrow C_6H_5-\underset{\diagdown\;CCl_2\;\diagup}{CH-N}-C_6H_5$$

Dihalomethylenes as Ring-Expanding Reagents. Ciamician and Dennstedt discovered that the reactions of pyrrylpotassium with chloroform and bromoform give 3-chloropyridine and 3-bromopyridine, respectively, in rather low yield (90). Although it seems clear that one function of the strongly basic pyrrylpotassium is to transform the haloform into dihalomethylene, it is not yet clear whether it is pyrrole or its anion that subsequently reacts with the dihalomethylene. The following is one of the possible mechanisms for the reaction:

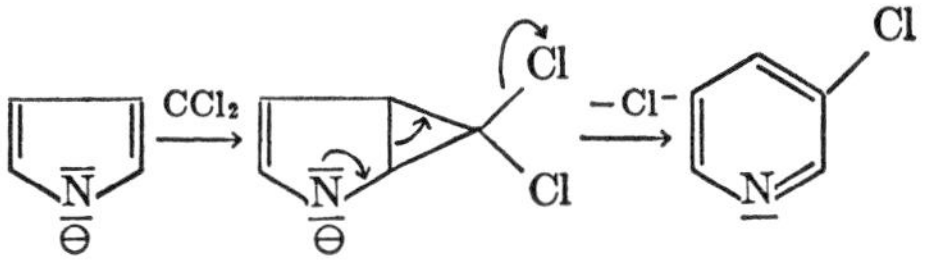

$$\text{pyrryl anion} \xrightarrow{CCl_2} \text{(dichlorocyclopropane-fused pyrryl anion)} \xrightarrow{-Cl^-} \text{3-chloropyridine}$$

Indoles give an analogous reaction, and in many cases the reaction is accompanied by the introduction of an aldehyde group into the pyrrole or indole ring, as in the Reimer-Tiemann reaction.

The first recognized use of dihalomethylenes for expanding carbocyclic rings appears to be due to Parham and Reiff who obtained 2-chloronaphthalene and apparently also a chloroazulene from the reaction of indenylsodium with chloroform (91). By use of indene, chloroform, and potassium *t*-butoxide it was possible to isolate the intermediate indene-dichloromethylene adduct, which was shown to be capable of first-order solvolysis yielding 2-chloronaphthalene (92). These observations may be explained in terms of the following reaction mechanism, but there is some evidence that the first carbonium ion shown is not a real intermediate but that the second carbonium ion shown is formed directly from the dihalocyclopropane by a concerted process.

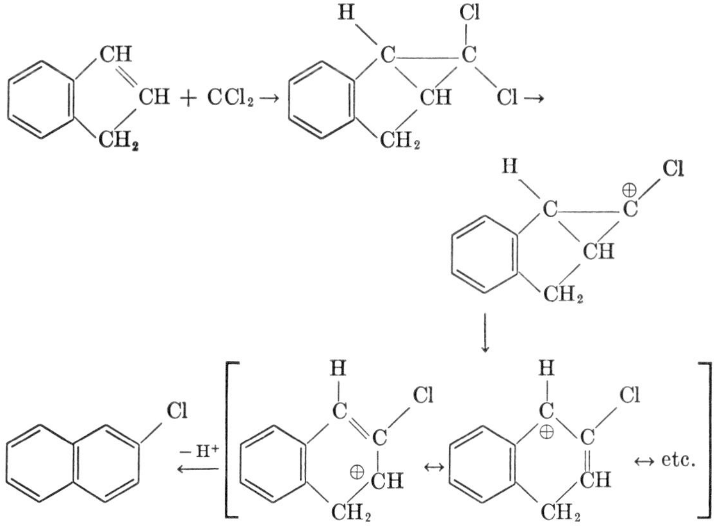

Although Parham and Reiff did not specifically identify the chloroazulene they obtained, there are other reports that show that dihalomethylenes may expand aromatic six-membered rings to seven-membered rings. Murray found that the reaction of anthracene with chloroform and potassium *t*-butoxide gives 10-chloro-5-*t*-butoxydibenzo[a,e]cycloheptatriene, perhaps by a path like the following (93):

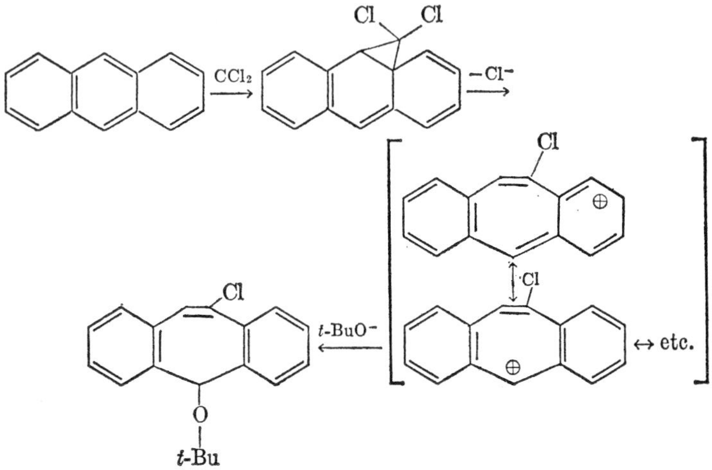

Cl Cl

CCl_2 →

$-Cl^-$ →

Cl

⊕

·Cl

Cl

t-BuO⁻ ←

↔ etc.

⊕

O

t-Bu

Parham, Bolon, and Schweizer found that the reaction of dichloromethylene with 9-methoxyanthracene or with either 1- or 2-methoxynaphthalene gave an adduct that lost methyl chloride on heating to yield a chlorobenztropone derivative (94).

Skell and Sandler made some interesting stereochemical observations in a study of the expansion of alicyclic rings (95). The reactions of silver nitrate with the adducts of cyclopentene with dibromomethylene, dichloromethylene, and bromochloromethylene (two geometric isomers as shown below) were all found to yield 2-halo-2-cyclohexene-1-ols.

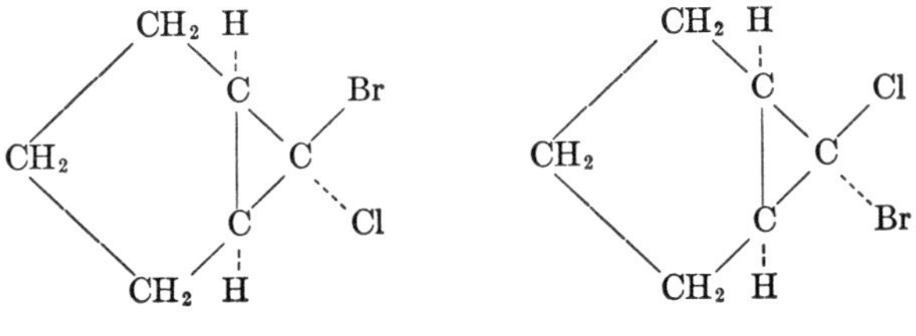

CH_2 H

C Br

CH_2 C

C Cl

CH_2 H

CH_2 H

C Cl

CH_2 C

C Br

CH_2 H

One of the isomeric bromochloromethylene adducts was found to lose only bromide ions and to do so at the same rate as the dibromomethylene adduct. The other lost only chloride ions and did so at the same rate as the dichloromethylene adduct. Thus there seems to be a clear stereochemical preference for the loss of one of the two halogens. It seems probable that it is the halogen *cis* to the five-membered ring that is lost more readily since the remaining halogen atom could then more closely approach coplanarity with the five nearest carbon atoms.

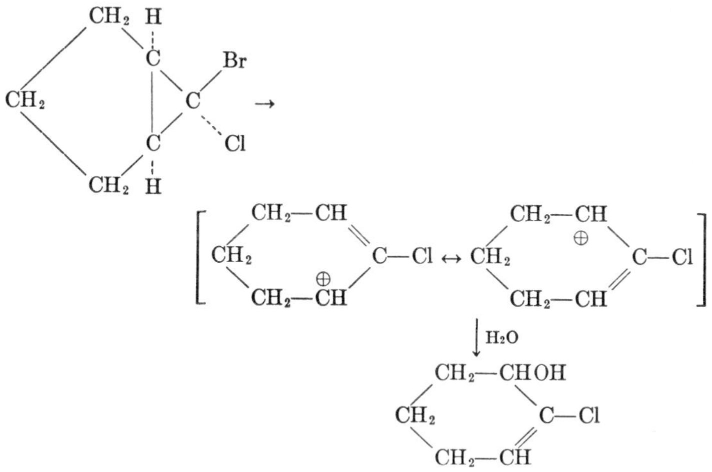

Vogel observed that the adduct of cyclooctatetraene and dichloromethylene rearranges not to a cyclononane but to a hydrogenated indane derivative (96). The first step of the reaction may be the valence-bond rearrangement reaction shown below.

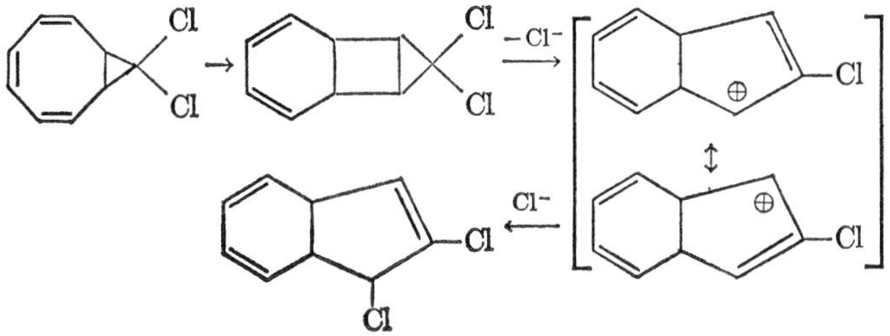

Neureiter discovered that the adduct of butadiene and dichloromethylene may be pyrolyzed to give a chlorocyclopentadiene (among other products) (97). As he pointed out, this reaction is probably initiated by a free-radical process, perhaps of the type:

$$CH_2{=}CH{-}\underset{\diagdown\ CCl_2\ \diagup}{CH{-}{-}CH_2} \rightarrow \left[CH_2{=}CH{-}\dot{C}H{-}\underset{\dot{C}Cl_2}{CH_2} \leftrightarrow \dot{C}H_2{-}CH{=}CH{-}\underset{\dot{C}Cl_2}{CH_2} \right]$$

$$\downarrow$$

$$\text{(ring: } CH{=}CH_2{-}CH\ \ CH{\diagup}\!\!{=}C(Cl)\text{)} \overset{?}{\leftrightarrows} \text{(ring: } CH{=}CH,\ CH,\ CH_2,\ C(Cl)\text{)} \xleftarrow{-HCl} \text{(ring: } CH{=}CH,\ CH_2,\ CH_2,\ CCl_2\text{)}$$

The pyrolysis of the cyclohexene-dichloromethylene adduct to give cycloheptatriene (and toluene) (98) may proceed by a similar but more complicated path.

The Reimer-Tiemann Reaction. The formation of *ortho* and *para* hydroxy aromatic aldehydes by the reaction of phenols with chloroform and alkali was discovered by Reimer and Tiemann and has been reviewed by Wynberg (99). From the observation that sodium phenoxide and chloroform alone are almost inert under conditions where the formation of *o*- and *p*-hydroxybenzaldehyde are rapid if excess sodium hydroxide is present, it seems clear that the reaction involves dichloromethylene (100). Since the phenoxide ion is the most reactive as well as the most abundant aromatic species present it is, no doubt, the species with which dichloromethylene reacts. Attack of dichloromethylene at the ortho carbon atom of the phenoxide ion, leading eventually to the formation of salicylaldehyde, may be initiated by the formation of the carbanion shown, or it may proceed by the alternate path indicated, in which a dichlorocyclopropane derivative is produced (99). The former intermediate lacks the strain inherent in a three-membered ring but the cyclopropane intermediate provides better resonance stabilization for the negative charge. Although it seems plausible that the subsequent steps shown above are among those followed in the formation of salicylaldehyde, this part of the reaction has not yet been studied in detail. Attack at the para position leads analogously to *p*-hydroxybenzaldehyde. Attack at oxygen may lead to the small amount of triphenyl orthoformate produced as a by-product, but it probably leads more commonly to carbon monoxide, formate ions, and the regeneration of phenoxide ions.

The "abnormal" Reimer-Tiemann reaction, in which dichloromethylcyclohexadienones are formed, also involves the reaction of dichloromethylene with a phenolate anion (101).

$$p\text{-}CH_3C_6H_4O^{\ominus} + CCl_2 \xrightarrow{H_2O} \text{4-methyl-4-(dichloromethyl)cyclohexa-2,5-dienone} \text{ and } \text{2-CHO-4-}CH_3C_6H_3O^{\ominus}$$

As would be expected, electron-donating substituents usually increase and electron-withdrawing substituents usually decrease the efficiency with which phenoxide ions combine with dichloromethylene. The yields of products obtained synthetically, however, may depend upon several other factors, such as the stability of the products to the reaction conditions. From the increase in the extent of ortho substitution that occurs when very high alkali concentrations and small cations are used, it seems likely that phenoxide-containing ion-pairs give more ortho substitution than do the free anions, perhaps because of an electrostatic effect.

Reactions of Dihalomethylenes with Amines. The formation of isocyanides from primary amines and chloroform is markedly catalyzed by strong bases and therefore does not involve a rate-controlling attack of the amine on chloroform. Since chloroform is known to be transformed to dichloromethylene under the reaction conditions the only plausible reaction mechanisms involve the combination of the amine with the dihalomethylene, followed by the loss of two molecules of hydrogen chloride.

$$CHCl_3 \xrightarrow{OH^-} CCl_2 \xrightarrow{RNH_2} R{-}\overset{\oplus}{N}H_2{-}\overset{\ominus}{C}Cl_2 \xrightarrow{-2HCl} R{-}N{=}C$$

Smith and Kalenda found that *N*-substituted formamides are also formed from at least some primary amines (102).

$$(CH_3)_2NCH_2CH_2CH_2NH_2 \xrightarrow[KOH]{CHCl_3} (CH_3)_2NCH_2CH_2CH_2NC \text{ and } (CH_3)_2NCH_2CH_2CH_2NHCHO$$

Saunders and Murray (103) showed that with secondary amines, where simple isocyanide formation is impossible, the formation of substituted formamides is rather general, and that with tertiary amines the dichloromethylene may insert itself between the nitrogen and a carbon atom, presumably via a Stevens-type rearrangement, to give, after hydrolysis, an *N*,*N*-disubstituted amide.

$$C_6H_5CH_2N(CH_3)_2 + CCl_2 \rightarrow C_6H_5CH_2\text{—}\overset{\oplus}{N}(CH_3)\text{—}CH_3 \text{ (with } \overset{\ominus}{C}Cl_2 \text{ on N)}$$

$$\downarrow$$

$$C_6H_5CH_2CON(CH_3)_2 \xleftarrow{H_2O} C_6H_5CH_2CCl_2N(CH_3)_2$$

In addition, tertiary amines may yield several other products, such as the *N*,*N*-disubstituted amides of α-chloroacids, whose mode of formation is less clear.

Clemens, Shropshire, and Emmons discovered that the sodium salt of *N*-methylaniline reacts with chlorodifluoromethane to give the first known orthoamide, which was shown to undergo a number of unusual reactions (104). In this case the haloform is transformed to difluoromethylene which then reacts with *N*-methylaniline or its conjugate base.

$$3C_6H_5N(CH_3)\text{—}Na + CHClF_2 \rightarrow CH(N(CH_3)C_6H_5)_3 + NaCl + 2NaF$$

Miscellaneous Reactions of Dihalomethylenes. Dihalomethylenes are, no doubt, capable of reacting with a wide variety of nucleophilic reagents. The reaction with halide ions, mentioned earlier in relation to the light it sheds on the mechanism of haloform hydrolysis, has also been used as a synthetic reaction. Bromochloroiodomethane may be prepared by the base-catalyzed reaction of dibromochloromethane with sodium iodide in methanol, and fluoroform can be produced by the reaction of chlorodifluoromethane with fluoride ions in basic solution (68).

$$CHBr_2Cl \xrightarrow{MeO^-} CBr_2Cl^- \xrightarrow{-Br^-} Br—C—Cl \xrightarrow{I^-} CBrClI^- \xrightarrow{MeOH} CHBrClI$$

Tertiary phosphines were found, almost simultaneously by two groups of workers (105, 106), to be capable of capturing dihalomethylenes to give stable phosphinemethylenes, which can then be used in a Wittig synthesis, to prepare 1,1-dihalo-olefins, e.g.,

$$(C_6H_5)_3P \xrightarrow[t\text{-BuOK}]{CHCl_3} (C_6H_5)_3P{=}CCl_2$$
$$(C_6H_5)_3P{=}CCl_2 + (C_6H_5)_2CO \rightarrow (C_6H_5)_2C{=}CCl_2 + (C_6H_5)_3PO$$

In the course of the extensive investigations of Parham and coworkers on the use of dihalomethylenes in ring-expansion reactions, an insertion into a C—H bond was observed in the reaction of a benzothiopyran (107).

$$\text{benzothiopyran} \xrightarrow[NaOMe]{Cl_3CCO_2Et} \text{2-}CHCl_2\text{-benzothiopyran} \text{ and } \text{4-}CHCl_2\text{-benzothiopyran}$$

From the nature of the two products shown it seems probable that the dichloromethylene did not attack the C—H bond but rather the sulfur atom or the resonance-stabilized conjugate base of the benzothiopyran.

The insertion reactions of dichloromethylene into benzylic C—H bonds, observed by Fields, cannot be explained on the basis of an intermediate benzyl anion since the dichloromethylene was generated by the decomposition of sodium trichloroacetate, a relatively weak base (108).

$$C_6H_5CH(CH_3)_2 \xrightarrow{Cl_3CCO_2Na} C_6H_5C(CH_3)_2CHCl_2$$

The selective attack at the benzyl position would be explained readily by a two-step reaction mechanism in which the dichloromethylene first abstracted the hydrogen atom to give intermediates that formed the carbon-carbon bond. The selectivity could also be explained by certain modifications of the one-step insertion mechanism.

Insertion at carbon-hydrogen bonds may be observed in higher yields when the dihalomethylene is generated from phenyltrihalomethylmercury compounds; even higher yields of products of insertion at silicon-hydrogen and germanium-hydrogen bonds have been obtained (108a).

4

Alkoxy-, Alkylthio-, and Monohalo-Methylenes

ALKOXY- (AND ARYLOXY-) METHYLENES

Monoalkoxymethylenes. Schöllkopf and coworkers found that the treatment of chloromethyl ethers with organolithium compounds in the presence of olefins leads to the formation of cyclopropane derivatives (109). This reaction must involve the metallation of the ether to give an intermediate organolithium compound that may react directly with the olefin or may, as shown below, decompose to a methylene, which then adds to the olefin.

$$ROCH_2Cl + RLi \rightarrow RO\overset{\displaystyle Li}{\overset{|}{C}}HCl \rightarrow RO{-}\ddot{C}{-}H$$

$$RO{-}\ddot{C}{-}H + (CH_3)_2C{=}CH_2 \rightarrow (CH_3)_2C\underset{\displaystyle \underset{RO\quad H}{\overset{C}{\diagup\diagdown}}}{\overline{\diagdown\qquad\diagup}}CH_2$$

With phenyl chloromethyl ether good yields of the cyclopropyl ethers may be obtained by use of n-butyllithium. With alkyl chloromethyl ethers, where the chlorine atom is more subject to nucleophilic displacement and the hydrogen atom to be replaced is less acidic, n-butyllithium yields largely alkyl pentyl ethers; the more basic and more hindered *t*-butyllithium, however, gives good yields of cyclopropyl alkyl ethers.

Alkoxyfluoromethylenes. The reaction of potassium isopropoxide with chlorodifluoromethane yields largely difluoromethyl isopropyl ether and triisopropyl orthoformate (110). From the rate of reaction and from data on the analogous reaction in methanol it seems clear that the chlorodifluoromethane first undergoes α-dehydrochlorination to give the intermediate difluoromethylene. Evidence as to the nature of the subsequent reactions of difluoromethylene comes from the observation that difluoromethyl isopropyl ether is inert to potassium isopropoxide in isopropyl alcohol under the reaction conditions. This observation proves that the triisopropyl orthoformate produced does not arise from the further reaction of the difluoromethyl ether. It may then be seen that although there remain a number of plausible mechanisms for the formation of the orthoester, one of which is shown below, all of these mechanisms involve the intermediate formation of isopropoxyfluoromethylene.

$$CF_2 \xrightarrow{i\text{-PrO}^-} i\text{-PrO}\overset{\ominus}{C}F_2 \rightarrow i\text{-PrO—C—F}$$

$$CF_2 \xrightarrow{i\text{-PrOH}} i\text{-PrOCHF}_2 \qquad i\text{-PrO}\overset{\ominus}{C}F_2 \rightarrow i\text{-PrOCHF}_2$$

$$i\text{-PrO—C—F} \xrightarrow{\text{several steps}} (i\text{-PrO})_3CH$$

It therefore seems reasonably well established that isopropoxyfluoromethylene is an intermediate in the reaction, and that in the reaction of chlorodifluoromethane with sodium methoxide in methanol, where analogous observations have been made, methoxyfluoromethylene is an intermediate.

The formation of alkyl difluoromethyl ethers as the principal products in reactions of difluoromethylene with alkoxide ions is no doubt related to the difficulty of displacing fluorine from carbon. When a more reactive halogen is present, as in chlorofluoromethylene, no dihalomethyl ether is observed. In the reaction of dichlorofluoromethane with potassium isopropoxide in isopropyl alcohol, for example, triisopropyl orthoformate is formed in high yield (111). Here, too, the reaction involves an alkoxyfluoromethylene intermediate, which is probably subsequently transformed to a dialkoxymethylene and then (perhaps in several steps) to the ortho ester.

$$\text{Cl—C—F} + i\text{-PrO}^- \rightarrow i\text{-PrO—C—F}$$

$$\downarrow i\text{-PrO}^-$$

$$(i\text{-PrO})_3\text{CH} \xleftarrow{i\text{-PrOH}} i\text{-PrO—C—OPr-}i$$

Apparently if the alkoxy group in an alkoxyfluoromethylene is relatively large the reaction of the methylene with additional alkoxide ions of the same type is hindered and the methylene dimerizes; at least, this seems to be the simplest explanation for the observation that the reaction of chlorodifluoromethane with potassium *t*-butoxide yields about 25 per cent *cis*- and *trans*-1,2-di-*t*-butoxy-1,2-difluoroethylene (112).

$$CF_2 + t\text{-BuO}^- \rightarrow t\text{-BuO—C—F} \rightarrow t\text{-BuOCF=CFOBu-}t$$

A small amount of the analogous isopropyl compound is found in the reaction with potassium isopropoxide but none of the analogous compounds were isolated when primary alkali-metal alkoxides were used.

Other Alkoxyhalomethylenes. So long as the reacting haloform contains a fluorine atom it will be the last halogen displaced; reactions with alkoxides will go through alkoxyfluoromethylenes (if they get as far as alkoxyhalomethylenes of any kind). When fluorine is absent, however, other alkoxyhalomethylenes are formed, and with quite a different result.

In 1855 Hermann observed that bromoform reacts with potassium hydroxide in aqueous ethanol to give ethylene. This olefin was also found by several other workers, some of whom used chloroform and some of whom used absolute ethanol. Propylene was obtained when *n*- or *iso*-propyl alcohol was used, and all the butyl alcohols, *t*-amyl alcohol, and cyclohexanol were found to yield the corresponding olefins (113). The observation by Skell and Starer that rearrangement often accompanies these dehydration reactions (e.g., isobutyl alcohol yields isobutylene, 1-butene, *cis*- and *trans*-2-butene, and methylcyclopropane) provides the best evidence for the mechanism of the reaction (114). These workers suggested that the alkoxybromo (or -chloro) methylene formed by the attack of an alkoxide ion on dihalomethylene subsequently ionizes to an ROC^+ cation, which is isoelectronic with an alkyldiazonium cation (RNN^+). In view of the ease with which the latter species loses the stable molecule nitrogen to yield a carbonium ion, it would be expected that the ROC^+ could similarly lose carbon monoxide to give a carbonium ion.

$$CHBr_3 \xrightarrow{RO^-} CBr_2 \xrightarrow{RO^-} R—\underline{\overline{O}}—\overline{C}—\underline{\overline{Br}}|$$

$$\downarrow -Br^-$$

$$R\oplus \xleftarrow{-CO} \left[R—\overset{\oplus}{\overline{O}}=\overline{C} \leftrightarrow R—\underline{\overline{O}}—\overline{C}\oplus \leftrightarrow \text{etc.} \right]$$

This carbonium ion may lose a β-hydrogen to give an unrearranged olefin or it may rearrange to another carbonium ion. The fact that about the same olefin mixtures are obtained as are produced by the reaction of the corresponding amines with nitrous acid is also evidence for the intermediacy of a carbonium ion. The ethers produced in the reaction probably also arise from the carbonium ion.

Methylene halides and ketones (when secondary alcohols are used) are also formed during the reaction. These may be formed by the action of dichloro-, dibromo-, and related methylenes as hydride-ion abstractors.

$$CH_3—\overset{|\overline{O}|^\ominus}{\underset{CH_3}{C}}—H \rightarrow \overset{Cl}{\underset{Cl}{C}}| \rightarrow CH_3\overset{O}{\underset{CH_3}{\overset{\|}{C}}} + \overset{\ominus}{C}HCl_2 \xrightarrow{ROH} CH_2Cl_2$$

Hydride-ion transfer is not observed with fluorine-containing dihalomethylenes because the fluorine atom destabilizes the required intermediate dihalomethyl anion and perhaps also because the fluorine substituent increases the tendency of the divalent carbon atom to coordinate with an oxygen atom and decreases the cationic character of the carbon atom by a resonance effect (111).

Alkoxychloromethylenes may be formed by the α-dehydrochlorination of alkyl dichloromethyl ethers. The reaction of dichloromethyl methyl ether with potassium isopropoxide in 25 per cent isopropyl alcohol–75 per cent benzene is 5.4 ± 2.2 times as fast as that of deuteriodichloromethyl methyl ether (115). From this primary deuterium kinetic isotope effect it seems clear that the hydrogen atom of the dichloromethyl group is being removed in the rate-controlling step of the reaction. This does not establish conclusively the intermediacy of a methylene in the reaction, but in view of the evidence for alkoxyhalomethylenes in other reactions it seems plausible that they are formed in the present case as well. The fact that the deuterium kinetic isotope effect is so much larger than that observed in the formation of trihalomethyl anions (56) suggests that the reaction is a concerted α-dehydrochlorination, not involving the intermediate formation of a carbanion.

$$i\text{-PrO}^- + CH_3OCHCl_2 \rightarrow i\text{-PrOH} + CH_3O\text{—C—Cl} + Cl^-$$

The concerted dehydrohalogenation of bromodifluoromethane, which shows an intermediate isotope effect (57), is probably a reaction with more carbanion character.

Dialkoxymethylenes. Scheibler's reports that dialkoxymethylenes can be prepared as pure compounds (carbon monoxide dialkyl acetals), stable at room temperature, were shown to be wrong by later workers. Dialkoxymethylenes may well be intermediates in the reactions of alkoxyhalomethylenes (including those formed from dihalomethylenes) with alcohols and alkoxide ions. Their intermediacy is difficult to prove, however, at least partly because they are not the first methylenes formed in these reactions.

It is possible that dimethoxymethylene is formed in the following transformation of a bicyclic compound to an aromatic compound studied by McBee, Idol, and Roberts (116).

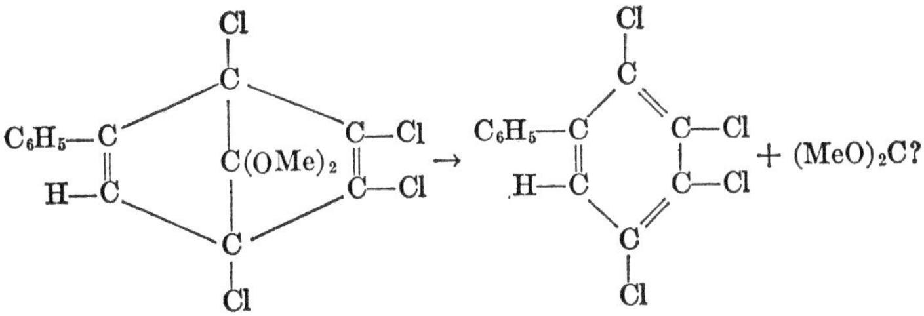

$$\text{C}_6\text{H}_5\text{—C}(=\text{CH})\text{ bicyclic adduct with bridge } \text{C(OMe)}_2\text{, bridgeheads C—Cl, and } \text{C(Cl)=C(Cl)} \rightarrow \text{C}_6\text{H}_5\text{-tetrachlorobenzene (C}_6\text{H}_5\text{—C, H—C, C—Cl, C—Cl, C—Cl, C—Cl)} + (\text{MeO})_2\text{C?}$$

No products derived from the bridge were isolated. Nevertheless, it is hoped that the mechanism of the reaction will be investigated; perhaps a method for the generation of a number of kinds of methylenes could be developed thereby. Corey and Winter have suggested that dialkoxymethylenes may be formed from trialkyl phosphites and thionocarbonates (116a).

ALKYLTHIOMETHYLENES

Phenylthiomethylene. Schöllkopf and Lehmann obtained for phenylthiomethylene evidence analogous to that described for monoalkoxymethylenes. The reaction of chloromethyl phenyl sulfide with butyllithium in the presence of olefins yields cyclopropyl phenyl sulfides (117).

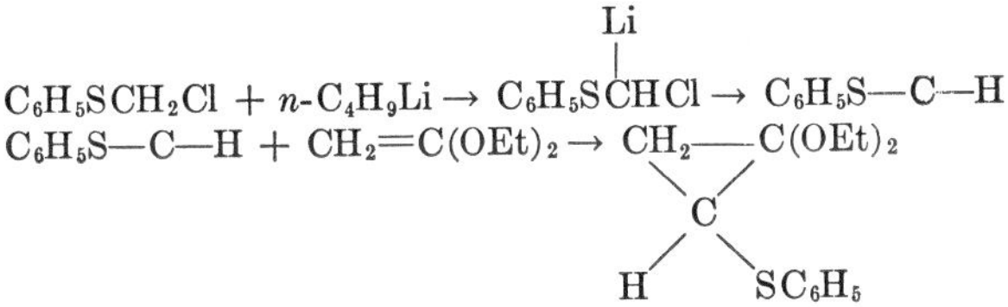

$$C_6H_5SCH_2Cl + n\text{-}C_4H_9Li \rightarrow C_6H_5S\overset{\displaystyle Li}{\overset{|}{C}}HCl \rightarrow C_6H_5S\text{—}\ddot{C}\text{—}H$$

$$C_6H_5S\text{—}\ddot{C}\text{—}H + CH_2\text{=}C(OEt)_2 \rightarrow \text{cyclopropane: } CH_2\text{—}C(OEt)_2\text{, }C(H)(SC_6H_5)$$

Alkylthiohalomethylenes. The evidence for the formation of alkylthiofluoromethylenes as reaction intermediates is essentially the same as that described earlier in this chapter for alkoxyfluoromethylenes. The sodium methoxide-catalyzed reaction of chlorodifluoromethane with sodium thiomethoxide in methanol yields difluoromethyl methyl sulfide and trimethyl orthothioformate as the principal organic products (118). Since the difluoromethyl methyl sulfide is essentially inert under the reaction conditions, the trimethyl orthothioformate must be formed by some mechanism involving the intermediate methylthiofluoromethylene, e.g.,

$$\begin{array}{ccccc} CF_2 + CH_3S^- & \rightarrow & CH_3S\overset{\ominus}{C}F_2 & \rightarrow & CH_3S{-}C{-}F \\ \downarrow{\scriptstyle CH_3SH} & \swarrow & & & \downarrow{\scriptstyle \text{several steps}} \\ CH_3SCHF_2 & & & & (CH_3S)_3CH \end{array}$$

Other alkylthiohalomethylenes are presumably formed in the reactions of other haloforms with mercaptide anions.

Bis(alkylthio)methylenes. It is reasonable to assume that bis(alkylthio)methylenes are intermediates in the formation of alkyl orthothioformates from haloforms, mercaptans, and strong base. The evidence for this is a bit indirect, however, and for some purposes such a method of generating bis(alkylthio)methylenes would be unsuitable. Thus if one wished to use certain methylene-capturing reagents to capture the bis(alkylthio)methylene produced from a haloform, one might well find that a dihalomethylene or an alkylthiohalomethylene (the first two methylenes formed in the reaction) was captured instead. The reaction of alkyl orthothioformates with very strong base is therefore of interest since there is evidence that bis(alkylthio)methylenes are formed, and if they are, they are certainly the first methylenes produced in the reaction.

This evidence was discovered by Arens and coworkers (119) and subsequently confirmed by other investigators (120), who, at the time, were unaware of Arens' prior work. In view of the fact that thioformals react with alkali-metal amides in liquid ammonia to give stable carbanions that can be alkylated by alkyl halides, it would certainly be expected that alkyl orthothioformates would also yield carbanions on reaction with amide ions. The formation of such tris(alkylthio)methyl anions is demonstrated by two observations. The addition of orthothioester to a colorless solution of potassium amide in liquid ammonia gives a green color similar to that observed with thioformals. The addition of methyl iodide to this green solution discharges the color with the formation of an orthothioacetate.

$$CH(SCH_3)_3 \xrightarrow{NH_2^-} \overset{\ominus}{C}(SCH_3)_3 \xrightarrow{CH_3I} CH_3C(SCH_3)_3$$

Upon standing for several hours the green solution containing carbanions becomes brown and a tetrakis(alkylthio)ethylene is formed. Since it is felt that the S_N1 or S_N2 displacement of an RS group from the orthothioformate is unlikely under the reaction conditions (where practically all the orthothioformate is present as its anion), it seems quite probable that the tetrakis(alkylthio)ethylenes are being formed from bis-(alkylthio)methylenes. They need not arise from the dimerization of the methylenes, however. They may instead be formed by the combination of the methylenes with the much more abundant tris(alkylthio)methyl anions to give pentakis-(alkylthio)ethyl anions, which then lose alkylthio anions.

$$\overset{\ominus}{C}(SCH_3)_3 \xrightarrow{-CH_3S^-} CH_3S{-}C{-}SCH_3$$

$$\downarrow \overset{\ominus}{C}(SCH_3)_3$$

$$(CH_3S)_2C{=}C(SCH_3)_2 \xleftarrow{-CH_3S^-} (CH_3S)_2\overset{\ominus}{C}{-}C(SCH_3)_3$$

MONOHALOMETHYLENES

Unsubstituted Monohalomethylenes. Evidence for the formation of chloromethylene as a reaction intermediate was published almost simultaneously by Volpin, Dulova, and Kursanov (121) and by Closs and Closs (122, 123). Kursanov and coworkers treated benzene with potassium *t*-butoxide and methylene chloride, methylene bromide, or methylene iodide (thus obtaining evidence for bromomethylene and iodomethylene also) and showed that tropilium *t*-butoxide was formed in yields ranging from 0.1 per cent to 1.4 per cent. A plausible mechanism for these reactions consists of addition of the halomethylene to the aromatic ring followed by ionization with rearrangement to a tropilium ion that then combines with a *t*-butoxide ion.

$$\text{C}_6\text{H}_6 + \text{H—C—X} \rightarrow \text{[norcaradiene-C(X)(H)]} \rightarrow \left[\text{C}_7\text{H}_7^+ \leftrightarrow \text{C}_7\text{H}_7^+ \leftrightarrow \text{etc.} \right] \xrightarrow{t\text{-BuO}^-} t\text{-BuO—C}_7\text{H}_7$$

Closs and Closs subsequently showed that methylcycloheptatriene is formed in 20 per cent yield by the action of methyllithium on methylene chloride in the presence of benzene (124), and that methyllithium, methylene chloride, and lithium phenoxide give 2-methyl-3,5-cycloheptadienone, presumably via the intermediate tropone (also formed, but only in about 0.1 per cent yield) (125).

$$\text{C}_6\text{H}_5\text{OLi} \xrightarrow{\text{H—C—Cl}} \text{tropone} \xrightarrow{\text{CH}_3\text{Li}} \text{(LiO, CH}_3\text{)-cycloheptatriene} \xrightarrow{\text{H}_2\text{O}} \text{2-methyl-3,5-cycloheptadienone}$$

The result obtained with lithium *o*-cresylate was analogous to that obtained with lithium phenoxide, but when the highly hindered 2,6-di-*t*-butylphenol was used 2,7-di-*t*-butyltropone was isolated in 23 per cent yield.

In their original study of chloromethylene Closs and Closs allowed various organolithium compounds to react with methylene chloride in the presence of olefins (122, 123). The intermediate chloromethylene reacts with the olefin to give a chlorocyclopropane and with the organolithium compound to give a terminal olefin.

$$n\text{-C}_4\text{H}_9\text{Li} + \text{CH}_2\text{Cl}_2 \rightarrow n\text{-C}_4\text{H}_{10} + \text{H—C—Cl} \xrightarrow{>\text{C=C}<} \text{—C}\overset{}{—}\text{C— (cyclopropane, C(H)(Cl))}$$

$$\text{H—C—Cl} \xrightarrow{n\text{-C}_4\text{H}_9\text{Li}} n\text{-C}_3\text{H}_7\text{CH=CH}_2$$

The olefin formation appears to involve the intermediacy of alkylmethylenes and is discussed further in Chapter 8. By use of *cis*- and *trans*-2-butene it was shown that chloromethylene adds to give cyclopropanes with retention of the geometric orientation of the reacting olefin. The nature of the products was further indicated by the fact that *cis*-2-butene gave two different chlorodimethylcyclopropanes while *trans*-2-butene gave but one.

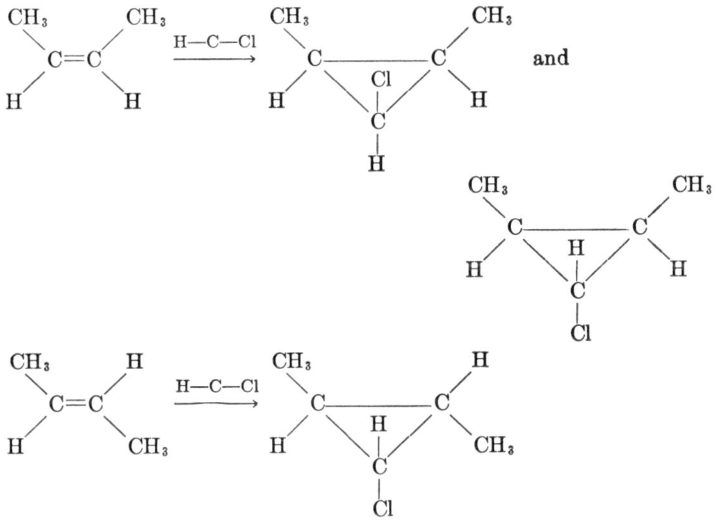

$$\begin{matrix} CH_3 & & CH_3 \\ & C{=}C & \\ H & & H \end{matrix} \xrightarrow{H-C-Cl} \begin{matrix} CH_3 & & & & CH_3 \\ & C & \text{—} & C & \\ H & & Cl & & H \\ & & C & & \\ & & H & & \end{matrix} \text{ and } \begin{matrix} CH_3 & & & & CH_3 \\ & C & \text{—} & C & \\ H & & H & & H \\ & & C & & \\ & & Cl & & \end{matrix}$$

$$\begin{matrix} CH_3 & & H \\ & C{=}C & \\ H & & CH_3 \end{matrix} \xrightarrow{H-C-Cl} \begin{matrix} CH_3 & & & & H \\ & C & \text{—} & C & \\ H & & H & & CH_3 \\ & & C & & \\ & & Cl & & \end{matrix}$$

Configurations were originally assigned to individual isomers, in the cases where two were formed, on the basis of the assumption that the isomer with the larger number of bulky substituents on one side of the cyclopropane ring would be formed in lower yield. Subsequent work showed that in at least some cases this is not so (123a).

The competition experiments of Closs and Schwartz, in which *n*-butyllithium was allowed to react with methylene chloride in the presence of various pairs of olefins, showed that electron-donating substituents increase the reactivity of olefins toward chloromethylene just as with dichloromethylene but that the magnitude of this increase in reactivity is less for chloromethylene (126). Thus, 2,3-dimethylbutene-2 is twelve times as reactive as pentene-1 toward chloromethylene but almost 400 times as reactive as pentene-1 toward dichloromethylene. The greater selectivity of dichloromethylene is probably due largely to its greater stability.

Chloromethylene has also been captured by triphenylphosphine, yielding products that may be used in the preparation of 1-chloroolefins via the Wittig reaction (106).

The α-dehydrochlorination of methylene chloride may be written either as a concerted process or as a two-step process involving the intermediate formation of a dichloromethyl anion and/or a dichloromethylmetal compound. As in the formation of dihalomethylenes the concerted mechanism should be favored by the formation of a more stable methylene or a more stable halide ion while the stepwise mechanism should be favored by substituents that stabilize an intermediate carbanion. Relative to chloroform, methylene chloride should yield a less stable methylene on dehydrochlorination but also a less stable intermediate carbanion. The two relevant factors thus act in opposite directions. In view of the surprisingly small difference in the acidities of diethyl thioformal and triethyl orthothioformate (both appear to be stronger acids than ammonia but weaker than triphenylmethane) (120), it seems that the methylene-stabilization factor should be more important and therefore that the reaction of methylene chloride with alkyllithiums should proceed through dichloromethyllithium, as Closs and Closs proposed (123).

Closs and Coyle found that the decomposition of chlorodiazomethane in the presence of olefins also gives chlorocyclopropane derivatives, but stereoisomers are formed in different ratios than when methylene chloride and an organolithium compound are used (126a). Furthermore, the decomposition of chlorodiazomethane in the presence of n-pentane leads to insertion products. Thus the intermediate formed from the diazo compound differs from that formed from methylene chloride, perhaps in excess energy, or perhaps because only dichloromethyllithium, and not free chloromethylene, is formed from methylene chloride.

Substituted Halomethylenes. Although intermediates of the type R—C—X have not been investigated as extensively as have unsubstituted monohalomethylenes, several species of this type have been suggested as intermediates in various reactions. The first such methylene for which significant evidence was reported is phenylchloromethylene. McElvain and Weyna found that the treatment of benzal chloride with sodium *t*-butoxide in the presence of ketene acetals yields derivatives of phenylchlorocyclopropanone acetals (87), e.g.,

$$(CH_3)_2C{=}C(OCH_3)_2 + C_6H_5CHCl_2 \xrightarrow{t\text{-BuONa}} \underset{\substack{\diagdown\ \diagup \\ C \\ \diagup\ \diagdown \\ C_6H_5 \quad Cl}}{(CH_3)_2C\text{———}C(OCH_3)_2}$$

Pyrolysis of some of these adducts yields α-phenylacrylates:

$$\underset{C_6H_5\text{—}C\text{—}Cl}{(CH_3)_2C\text{———}C(OCH_3)_2} \xrightarrow{-Cl^-}$$

$$\underset{\substack{\diagdown\ \diagup \\ C\oplus \\ | \\ C_6H_5}}{(CH_3)_2C\text{———}C(OCH_3)_2} \rightarrow \left[\underset{\substack{\diagdown\ \diagup\!\!\diagup \\ C \\ | \\ C_6H_5}}{(CH_3)_2\overset{\oplus}{C} \qquad C(OCH_3)_2} \longleftrightarrow \underset{\substack{\diagdown\!\!\diagdown\ \diagup \\ C \\ | \\ C_6H_5}}{(CH_3)_2C \qquad \overset{\oplus}{C}(OCH_3)_2} \right]$$

$$\xrightarrow{Cl^-} (CH_3)_2C{=}\underset{C_6H_5}{\underset{|}{C}}\text{—}\overset{O}{\overset{\|}{C}}OCH_3 + CH_3Cl$$

Breslow, Haynie, and Mirra used this reaction to prepare the first known derivative of cyclopropenone from benzal chloride, potassium *t*-butoxide, and phenyl ketene dimethyl acetal, probably by the following mechanism (127).

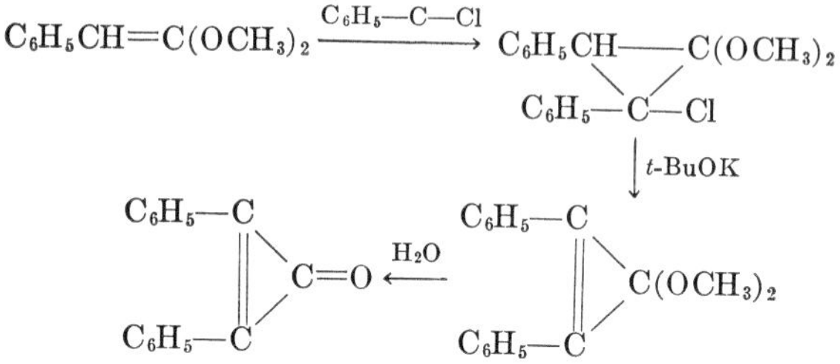

$$C_6H_5CH{=}C(OCH_3)_2 \xrightarrow{C_6H_5-\ddot{C}-Cl} C_6H_5CH\!-\!\!-\!C(OCH_3)_2 \text{ (cyclopropane ring with } C_6H_5-C-Cl)$$

$$\downarrow t\text{-BuOK}$$

$$\text{(cyclopropenone: } C_6H_5-C{=}C-C_6H_5 \text{ ring with } C{=}O) \xleftarrow{H_2O} \text{(cyclopropene: } C_6H_5-C{=}C-C_6H_5 \text{ ring with } C(OCH_3)_2)$$

Kirmse described evidence that 1-chloroethylidene is generated by the action of alkyllithium compounds on ethylidene chloride (128).

To explain the rearranged structures of the haloolefins reported as the major products in the pyrolysis of two polyhaloethyltrichlorosilanes Haszeldine and Young suggested that intermediate methylenes may form and rearrange, as shown below (129).

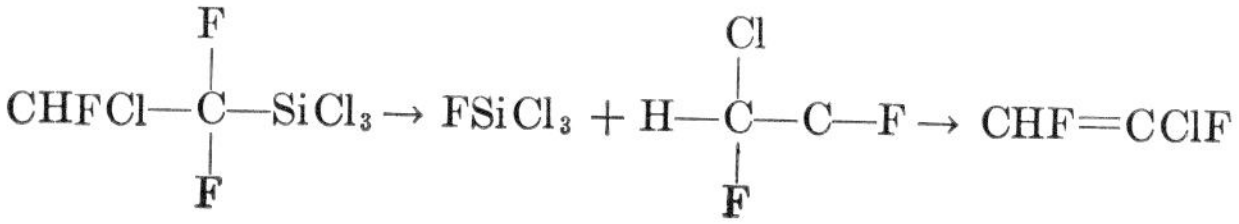

$$CHFCl-CF_2-SiCl_3 \rightarrow FSiCl_3 + H-CClF-\ddot{C}-F \rightarrow CHF{=}CClF$$

5

Double-Bonded Divalent Carbon

CARBON MONOXIDE

Carbon monoxide, the only really common chemical that is a divalent carbon derivative, has been the subject of a great deal of research and its chemistry has been discussed in a number of reviews and books. Only a few of the chemical and physical properties of carbon monoxide will be mentioned here, these being chosen to provide points for comparison with other divalent carbon derivatives.

Carbon monoxide is isoelectronic with molecular nitrogen and melts and boils only about three degrees higher. Like nitrogen, carbon monoxide is a relatively stable compound with a large heat of formation from the atoms. The divalent character of the carbon atom, however, may be seen in the reactivity (relative to nitrogen) of carbon monoxide toward a variety of reagents. Carbon monoxide reacts with sodium hydroxide, at a moderate rate even in aqueous solution at room temperature, to give sodium formate. Analogously, sodium alkoxides catalyze the addition of alcohols to carbon monoxide to give alkyl formates, a reaction that may proceed as follows:

$$\mathrm{CO + RO^- \rightleftharpoons RO{-}\overset{\ominus}{\overline{C}}{=}O}$$

$$\mathrm{RO{-}\overset{\ominus}{\overline{C}}{=}O + ROH \rightleftharpoons ROCHO + RO^-}$$

The reaction, as shown, is reversible and is carried out under a high pressure of carbon monoxide. At one atmosphere alkyl formates may be decomposed by alkali-metal alkoxides to give carbon monoxide and alcohol.

Carbon monoxide is also capable of reacting with Brønsted and Lewis acids, perhaps with the formation of an intermediate acyl cation, as in the following suggested mechanism for the aluminum chloride-catalyzed addition of carbon tetrachloride to carbon monoxide.

$$CCl_4 + AlCl_3 \rightleftharpoons CCl_3^+ + AlCl_4^-$$

$$CCl_3^+ + CO \rightleftharpoons Cl_3C{-}\overset{\oplus}{C}{=}O$$

$$Cl_3C{-}\overset{\oplus}{C}{=}O + AlCl_4^- \rightleftharpoons Cl_3C{-}COCl + AlCl_3$$

Carbon monoxide may be eliminated from most carboxylic acids, RCO_2H, by the action of strongly acidic catalysts, if R forms a fairly stable cation. Thus acid-catalyzed decarbonylation is a rather general reaction for α-hydroxy-, trialkylacetic-, and similar acids.

$$R{-}\underset{\displaystyle OH}{\underset{|}{C}H}{-}CO_2H \xrightarrow{H^+} R{-}\underset{\displaystyle OH}{\underset{|}{C}H}{-}\overset{\displaystyle O}{\overset{\|}{C}}{-}OH_2^+ \xrightarrow{-H_2O} R{-}\underset{\displaystyle OH}{\underset{|}{C}H}{-}\overset{\displaystyle O}{\overset{\|}{C}}\oplus$$

$$\downarrow -CO$$

$$RCHO \xleftarrow{-H^+} R{-}\underset{\displaystyle OH}{\underset{|}{C}H}\oplus$$

A number of reactions of this type have been found to be reversible.

Several free-radical reactions involving carbon monoxide illustrate the particular stability of acyl radicals. Most aliphatic aldehydes, when heated with di-*t*-butyl peroxide, undergo decarbonylation.

$$RCHO + t\text{-}BuO\cdot \rightarrow R{-}\dot{C}{=}O + t\text{-}BuOH$$

$$R{-}\dot{C}{=}O \rightarrow R\cdot + CO$$

$$R\cdot + RCHO \rightarrow RH + R{-}\dot{C}{=}O$$

Acyl radicals are presumably formed in the reported radical copolymerization of carbon monoxide and ethylene.

$$R\cdot \xrightarrow{CO} RCO\cdot \xrightarrow{C_2H_4} RCOCH_2CH_2\cdot \xrightarrow{CO} RCOCH_2CH_2CO\cdot \xrightarrow{\text{etc.}}$$

In addition to the participation of carbon monoxide in some of the syntheses and decomposition reactions of organic molecules with classical types of structures, the divalent character of the carbon atom in carbon monoxide causes reactions with metals to give carbonyls, whose valence-bond structures are not so well understood.

$$Fe + 5CO \rightarrow Fe(CO)_5 \qquad \text{b.p. } 102^\circ$$

ISOCYANIDES

Isocyanides, also known as isonitriles or carbylamines, comprise the only *class* of derivatives of divalent carbon known as stable, isolable compounds. Their chemistry has been reviewed by Bruylants (130) and will be outlined only briefly here.

The preparation of isocyanides by the reaction of primary amines with haloforms in the presence of strong base was described in Chapter 3. The reaction of potassium cyanide with alkyl halides often gives alkyl cyanides in good yield but never more than very small amounts of isocyanides. However, the reaction of alkyl halides with the cyanides of certain heavier metals, preferably silver, often gives rather good yields of alkyl isocyanides.

Since the R—N= substituent should be a weaker electron-withdrawer and a stronger electron-donor than the O= substituent, it is not surprising that isocyanides are generally less reactive toward bases but more reactive toward acids than is carbon monoxide. Thus isocyanides are stable to alkali even at 100° but are readily hydrolyzed by dilute aqueous acid to give formic acid and primary amines.

$$\left[\begin{array}{c} R{-}\bar{N}{=}C| \\ \updownarrow \\ R{-}\overset{\oplus}{N}{\equiv}\overset{\ominus}{C}| \end{array}\right] \xrightarrow{H^+} \left[\begin{array}{c} R{-}\bar{N}{=}\overset{\oplus}{C}{-}H \\ \updownarrow \\ R{-}\overset{\oplus}{N}{\equiv}C{-}H \end{array}\right] \xrightarrow{H_2O} R{-}\bar{N}{=}\underset{}{\overset{\overset{\oplus}{O}H_2}{C}}{-}H$$

$$\downarrow$$

$$RNH_2 + HCO_2H \xleftarrow{H_2O} R{-}NH{-}\overset{O}{\overset{\|}{C}}H$$

Isocyanides react violently with glacial acetic acid giving acetic anhydride and *N*-substituted formamides, probably by a mechanism somewhat like the following:

$$R{-}N{=}C + HOAc \rightarrow R{-}N{=}\overset{\oplus}{C}{-}H + AcO^-$$

$$\downarrow$$

$$Ac_2O + RNHCHO \xleftarrow{HOAc} R{-}N{=}CHOAc$$

They also add chlorine, acyl halides, hydrogen sulfide, and many other reagents.

$$\underset{\displaystyle R{-}C{=}O}{R{-}N{=}C{-}Cl} \xleftarrow{RCOCl} R{-}N{=}C \xrightarrow{H_2S} R{-}NH{-}\overset{S}{\overset{\|}{C}}H$$

$$R{-}N{=}C \xrightarrow{Cl_2} R{-}N{=}CCl_2$$

With certain metals complexes are formed that resemble metal carbonyls.

$$Ni(CO)_4 + 4C_6H_5NC \rightarrow (C_6H_5NC)_4Ni + 4CO$$

INTERMEDIATES OF THE TYPE >C=C

Since the only stable derivatives of divalent carbon known are those in which the divalent carbon is attached by a double bond to oxygen or nitrogen, it is not surprising that the possible existence of species with divalent carbon attached by a double bond to carbon has also been investigated. Since carbon is less electronegative than nitrogen or oxygen the $R_2C{=}$ substituent is not as capable of electron withdrawal as is RN= or O=. Since the R_2C= substituent lacks unshared electron pairs it is not as capable of resonance-electron-donation as is RN= or O=. These facts may help explain why species of the type R_2C=C are not nearly so stable as carbon monoxide or organic isocyanides.

The α-Dehydrohalogenation of 2,2-Diarylvinyl Halides. In spite of the relative stability that might be expected for derivatives of double-bonded divalent carbon, the transformation of 2,2-diarylvinyl halides to diarylacetylenes, a reaction for which a divalent carbon intermediate has been suggested, almost undoubtedly does not ordinarily involve such an intermediate. Bothner-By showed that the reactions of *cis*- and of *trans*-1-*p*-bromophenyl-1-phenyl-2-bromoethylene-1-^{14}C with potassium *t*-butoxide both proceed with about 90 per cent migration of the aryl group *trans* to the vinyl bromine atom (131).

$$(p\text{-BrC}_6\text{H}_4)(\text{C}_6\text{H}_5)\text{C}^*\text{=CHBr} + t\text{-BuO}^- \rightleftharpoons (p\text{-BrC}_6\text{H}_4)(\text{C}_6\text{H}_5)\text{C}^*\text{=}\overset{\ominus}{\text{C}}\text{Br} + t\text{-BuOH}$$

$$\xrightarrow{-\text{Br}^-} \text{C}_6\text{H}_5\overset{*}{\text{C}}\equiv\text{CC}_6\text{H}_4\text{Br-}p$$

Curtin, Flynn, and Nystrom obtained similar results in the reaction of *cis*- and *trans*-1-phenyl-1-*p*-chlorophenyl-2-bromoethylene-1-^{14}C with butyllithium in ether (132). The stereospecificity of this rearrangement shows that no free 2,2-diarylvinylidene was formed as an intermediate.

$$(\text{Ar}')(\text{Ar})\text{C=CHBr} \not\longrightarrow (\text{Ar}')(\text{Ar})\text{C=C|}$$

Since the same vinylidene intermediate would be formed from the *cis* as from the *trans* reactant, the same products would be formed also and there would have been no preference for migration by the aryl group *trans* to the bromine. The observation by Pritchard and Bothner-By that in the case of 2,2-diphenylvinyl bromide the hydrogen α to the bromine atom undergoes base-catalyzed exchange with the solvent at a rate much faster than the formation of diarylacetylene shows that the reaction is initiated by the reversible formation of a carbanion, as indicated in the reaction scheme given (133). This carbanion then undergoes a rate-controlling rearrangement that is analogous to the rate-controlling step of the Beckmann rearrangement. The lack of *complete* stereospecificity in the reaction could be explained in terms of an alternate mechanistic path involving an intermediate diphenylvinylidene, but an equally plausible explanation is that the intermediate carbanions occasionally undergo *cis-trans* isomerization.

The α-Dehydrohalogenation of 9-Halomethylenefluorenes. One can regard 9-bromomethylenefluorene as a 2,2-diarylvinyl bromide whose aryl groups are tied together so that migration of either one is hindered, and the possible acetylenic product of α-dehydrobromination with rearrangement cannot attain the linear configuration required for maximum stability by an acetylene.

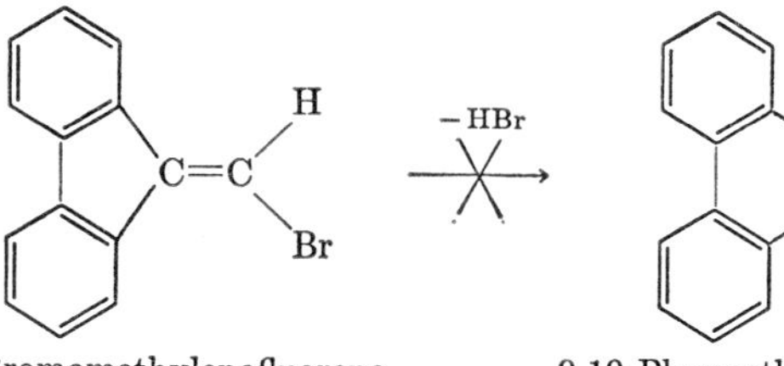

$-HBr$ (reaction does not proceed)

9-Bromomethylenefluorene → 9,10-Phenanthryne

As Hauser and Lednicer pointed out this is, no doubt, the reason why 9-bromomethylenefluorene does not react with base to give a phenanthrene derivative (134). With potassium amide in liquid ammonia it gives instead as much as 95 per cent of a cumulene (1,4-dibiphenylenebutatriene). As shown below, several plausible paths may be written for this reaction, one involving successive carbanion formation, nucleophilic attack on unreacted bromide, and dehydrobromination, and another involving α-dehydrobromination to give 2-biphenylenevinylidene, which may yield the cumulene either by dimerization or by combination with the initial carbanion formed from the reactant followed by loss of bromide ion.

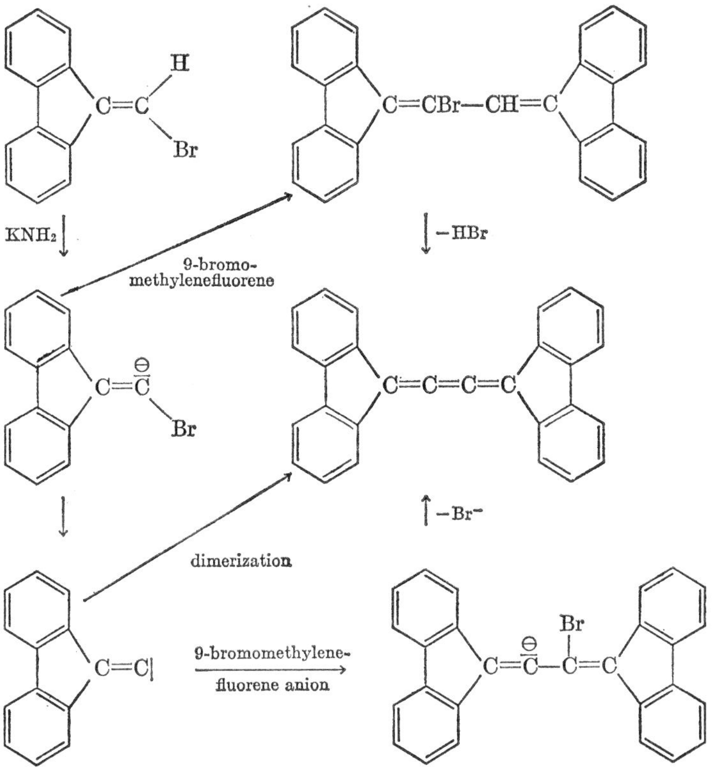

KNH_2

9-bromo-
methylenefluorene

$-HBr$

$-Br^-$

dimerization

9-bromomethylene-
fluorene anion

Curtin and Richardson found that the reaction of 9-bromomethylenefluorene with phenyllithium in ether also gives 1,4-dibiphenylenebutatriene (135). For this reaction, mechanisms may be written analogous to those shown above, except that the carbanions, which are probably present in ion-pairs in liquid ammonia, are more reasonably replaced by organolithium compounds with significant covalent character in the C—Li bonds. As Curtin and Richardson point out, the alkylation-dehydrohalogenation mechanism seems less probable than the mechanisms involving the intermediate formation of 2-biphenylenevinylidene. More evidence is needed, however, to establish this intermediate very definitely.

Intermediates of the Type $\overset{\diagdown}{-}C{=}C{=}C$. Hennion and Maloney found that the solvolysis of both 3-chloro-3-methyl-1-butyne and 1-chloro-3-methyl-1,2-butadiene is greatly accelerated by base and that both reactants yield 3-ethoxy-3-methyl-1-butyne in 80 per cent ethanol-water (136). In view of the well-known acidity of terminal acetylenes, the suggestion by Hennion and Maloney that the reaction of the acetylenic chloride involves the formation of a carbanion followed by loss of a halide ion to give a reactive intermediate divalent-carbon derivative seems to have a firm basis. The same intermediate could be formed from the chloroallene, as shown below.

$$CH_3{-}\underset{Cl}{\overset{CH_3}{\underset{|}{\overset{|}{C}}}}{-}C{\equiv}CH \overset{EtO^-}{\rightleftharpoons} CH_3{-}\underset{Cl}{\overset{CH_3}{\underset{|}{\overset{|}{C}}}}{-}C{\equiv}\overset{\ominus}{C|} \xrightarrow{-Cl^-} \left[(CH_3)_2\overset{\oplus}{C}{-}C{\equiv}\overset{\ominus}{C|} \longleftrightarrow (CH_3)_2C{=}C{=}C| \right]$$

$$(CH_3)_2C{=}C{=}C\begin{smallmatrix}H\\Cl\end{smallmatrix} \overset{EtO^-}{\rightleftharpoons} (CH_3)_2C{=}C{=}\overset{\ominus}{\overline{C}}{-}Cl \xrightarrow{-Cl^-} \left[(CH_3)_2C{=}C{=}C| \right]$$

$$\left[\ \right] \xrightarrow{EtO^-} CH_3{-}\underset{OEt}{\overset{CH_3}{\underset{|}{\overset{|}{C}}}}{-}C{\equiv}\overset{\ominus}{C|} \xrightarrow{EtOH} CH_3{-}\underset{OEt}{\overset{CH_3}{\underset{|}{\overset{|}{C}}}}{-}C{\equiv}CH$$

As Hennion and Nelson pointed out (137), further evidence for the proposed reaction mechanism may be found in the fact that second-order substitution no longer occurs when the reacting acetylenic chloride lacks a terminal triple bond, as in the case of 4-chloro-4-methyl-2-pentyne, for example. In addition, Hennion and Teach showed that unlike most tertiary aliphatic halides, which react with sodamide to give olefins, 3-chloro-3-methyl-1-butyne reacts with sodamide or even with the sodium salt of acetylene in liquid ammonia to give the corresponding primary amine (138).

$$CH_3{-}C(CH_3)(Cl){-}C{\equiv}CH \xrightarrow[NH_3]{\text{strong base}} CH_3{-}C(CH_3)(NH_2){-}C{\equiv}CH$$

Hartzler was able to capture the intermediate methylene with several different olefins (139). The reaction of 3-chloro-3-methyl-1-butyne with alcohol-free potassium *t*-butoxide in the presence of *cis*-2-butene, for example, appears to proceed as follows:

$$CH_3{-}C(CH_3)(Cl){-}C{\equiv}CH \xrightarrow{t\text{-BuOK}} \left[(CH_3)_2\overset{\oplus}{C}{-}C{\equiv}\overset{\ominus}{C}| \longleftrightarrow (CH_3)_2C{=}C{=}C \right]$$

$$\xrightarrow{cis\text{-2-butene}} \text{cyclopropane: } (CH_3)(H)C{-}C(H)(CH_3) \text{ ring closed by } C{=}C{=}C(CH_3)_2$$

The addition is apparently stereospecific since a different 1-(2-methylpropenylidene)-2,3-dimethylcyclopropane is obtained when *trans*-2-butene is used. The intermediate methylenes behave as electrophilic reagents toward olefins, whose relative reactivities are increased by the presence of alkyl groups on the carbon atoms of the double bond. The reactivity sequence, 2,3-dimethyl-2-butene > 2-methyl-1-butene $\sim$ 2-methyl-2-butene > cyclohexene > 1-hexene, was observed, with relative reactivities being of about the same magnitude as those observed for reaction with dibromomethylene.

The acetates of diarylethynyl carbinols react with strong bases to give 1,1,6,6-tetraarylhexapentaenes, which can be written as arising from the dimerization of intermediate diarylvinylidenemethylenes, but which, as Hartzler pointed out, probably are formed by the combination of the methylene with an intermediate carbanion with the loss of an acetate ion as shown below.

$$\underset{\displaystyle\text{OAc}}{\text{Ar}_2\text{C}}\text{—C}\equiv\text{CH} \xrightarrow{-\text{H}^+} \underset{\displaystyle\text{OAc}}{\text{Ar}_2\text{C}}\text{—C}\equiv\text{C}|^{\ominus} \xrightarrow{-\text{OAc}^-} \text{Ar}_2\text{C}{=}\text{C}{=}\text{C}$$

$$\downarrow \; \underset{\displaystyle\text{OAc}}{\text{Ar}_2\text{C}}\text{—C}\equiv\text{C}|^{\ominus}$$

$$\text{Ar}_2\text{C}{=}\text{C}{=}\text{C}{=}\text{C}{=}\text{C}{=}\text{CAr}_2 \xleftarrow{-\text{OAc}^-} \underset{\displaystyle\text{OAc}}{\text{Ar}_2\text{C}}\text{—C}\equiv\text{C—}\overset{\ominus}{\text{C}}{=}\text{C}{=}\text{CAr}_2$$

The formation of cyclopropane derivatives in the reactions of certain compounds in the presence of olefins provides strong evidence for the intermediacy of methylenes. The evidence is not so strong, however, as that provided by kinetic studies in favorable cases. Shiner and Wilson obtained kinetic evidence for the intermediacy of a methylene in the reaction of 3-bromo-3-methyl-1-butyne with base in 80 per cent aqueous ethanol. They found that base-catalyzed deuterium exchange of the acetylenic hydrogen atom is much faster than base-catalyzed decomposition of the organic bromide, showing that the rate-controlling step in the reaction comes after the initial carbanion formation (140). The most convincing evidence was the observation of a mass law effect (cf. p. 38). The reaction of the organic bromide with base in the presence of 0.22 *M* sodium bromide was about 26 per cent slower than in the presence of 0.22 *M* sodium perchlorate or sodium nitrate, salts whose anions are relatively weakly nucleophilic. Apparently the bromide ions capture the intermediate methylenes.

$$\left[\begin{array}{c}(CH_3)_2C{=}C{=}C| \\ \updownarrow \\ (CH_3)_2\overset{\oplus}{C}{-}C{\equiv}C|^{\ominus}\end{array}\right] \xrightarrow{Br^-} \begin{array}{l}(CH_3)_2C{=}C{=}\overset{\ominus}{\underline{C}}{-}Br \rightarrow (CH_3)_2C{=}C{=}CHBr \quad \text{I} \\ \\ (CH_3)_2\underset{Br}{\underset{|}{C}}{-}C{\equiv}C|^{\ominus} \rightarrow (CH_3)_2\underset{Br}{\underset{|}{C}}{-}C{\equiv}CH \quad \text{II}\end{array}$$

In order to explain the fall in the second-order rate constants that is observed as the reaction proceeds, especially in the presence of large concentrations of bromide ions, it is necessary to assume that a large fraction of the bromide ions that combine with the intermediate methylenes do so to give the allenic bromide (I), which is considerably less reactive than the acetylenic bromide (II).

Alkaline Decomposition of 3-Nitroso-2-oxazolidones. Newman and coworkers obtained a number of interesting products from the treatment of 3-nitrosooxazolidones with alkali (141). They suggest for most of these reactions a mechanism that involves an intermediate unsaturated carbonium ion and that seems to explain all the experimental observations adequately. In the case of the transformation of 5,5-diphenyl-3-nitroso-2-oxazolidone to diphenylacetylene this mechanism may be written:

$$(C_6H_5)_2\overset{}{C}\!-\!CH_2\!-\!N(NO)\!-\!CO\!-\!O\ (\text{ring}) \xrightarrow[H_2O]{OH^-} (C_6H_5)_2C(OCO_2H)CH_2NHNO \rightarrow (C_6H_5)_2C(OCO_2^-)CH_2N{=}NOH$$

$$\xrightarrow{OH^-} (C_6H_5)_2C{=}C(H)N_2OH \xrightarrow[-N_2]{-OH^-} (C_6H_5)_2C{=}\overset{\oplus}{C}H \rightarrow C_6H_5\overset{\oplus}{C}{=}CHC_6H_5 \xrightarrow{-H^+} C_6H_5C{\equiv}CC_6H_5$$

Starting with the unsaturated diazohydroxide in the mechanism above, several alternative mechanisms may also be written, some involving intermediate methylenes, e.g.,

$$(C_6H_5)_2C{=}C(H)N{=}NOH \xrightarrow{OH^-} (C_6H_5)_2C{=}C{=}N{=}\overline{N}| \xrightarrow{-N_2} (C_6H_5)_2C{=}C| \rightarrow C_6H_5C{\equiv}CC_6H_5$$

There does not seem to be enough evidence, however, to support any one of the various possible mechanisms very strongly relative to the others.

The Decomposition of Carbon Suboxide. The decomposition of carbon suboxide is a particularly interesting reaction that appears to involve the simultaneous formation of two double-bonded divalent carbon derivatives. Bayes observed that the photolysis of carbon suboxide in the presence of ethylene yields allene and about twice as much carbon monoxide (142). The decomposition of photoexcited carbon suboxide apparently involves simple cleavage of the carbon-

carbon double bond, just as in the photolysis of other ketenes. The 2-oxovinylidene (C_2O) thus formed may add to ethylene to give an intermediate ketene with so much excess energy that it immediately loses carbon monoxide. This loss of carbon monoxide gives cyclopropylidene, which is known to rearrange rapidly to allene.

$$O{=}C{=}C{=}C{=}O \xrightarrow{h\nu} CO + C{=}C{=}O \xrightarrow{C_2H_4} \begin{matrix} CH_2 \\ | \quad \diagdown \\ | \quad\quad C{=}C{=}O^* \\ | \quad \diagup \\ CH_2 \end{matrix}$$

$$\downarrow$$

$$CH_2{=}C{=}CH_2 \leftarrow \begin{matrix} CH_2 \\ | \quad \diagdown \\ | \quad\quad C + CO \\ | \quad \diagup \\ CH_2 \end{matrix}$$

As an alternative mechanism it may be suggested that atomic carbon is formed either directly in the photolysis reaction or subsequently by the decomposition of C_2O, and it is the atomic carbon that attacks ethylene.

$$CH_2{=}CH_2 + C \rightarrow \begin{matrix} CH_2 \text{---} CH_2 \\ \diagdown \;\; \diagup \\ C \end{matrix} \rightarrow CH_2{=}C{=}CH_2$$

However, MacKay, Polak, Rosenberg, and Wolfgang showed that atomic carbon is very probably not an intermediate in the reaction (143). It seems doubtful that light of the wave length used (2537 Å) contains enough energy to break both of the carbon-carbon double bonds in carbon suboxide. Even the subsequent decomposition of C_2O appears to require too much energy to proceed at an appreciable rate under the conditions used. Even more convincingly, atomic carbon (recoil ^{11}C from a nuclear reaction) reacts with ethylene to give about twice as much acetylene as allene. Since no such amount of acetylene seems to have been formed in Bayes' experiments, atomic carbon must not have been a reaction intermediate. The possibility that the large yield of acetylene obtained by MacKay and coworkers was due to the possession of much more excess energy by their atomic carbon was ruled out by experiments in which a twenty-fold excess of neon was present to bring about the collisional deactivation of the "hot" atomic carbon. Under these conditions the ratio of acetylene to allene did not change significantly.

Divalent Carbon in the Molecule C_3. The molecule C_3, first detected in comets and then terrestrially in carbon vapor, appears to contain two divalent carbon atoms. Skell and Wescott evaporated elemental carbon under a high vacuum onto a liquid nitrogen-cooled surface covered with olefins and obtained allenes containing two three-membered rings (143a).

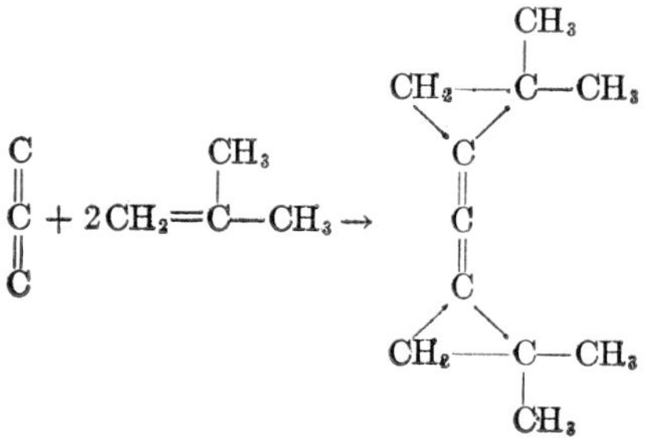

$$C{=}C{=}C + 2CH_2{=}C(CH_3){-}CH_3 \rightarrow (CH_3)_2C{<}\overset{CH_2}{|}{>}C{=}C{=}C{<}\overset{CH_2}{|}{>}C(CH_3)_2$$

6

Formation of Miscellaneous Methylenes by Base-Induced α-Elimination Reactions

α-ELIMINATION REACTIONS OF BENZYL HALIDES, ETHERS, AND SULFONIUM SALTS

The Reaction of Nitrobenzyl Halides and Sulfonium Ions with Base. Michael and later Bergmann and Hervey suggested that the reaction of *o*- or *p*-nitrobenzyl halides with sodium ethoxide to give a dinitrostilbene is initiated by an α-dehydrohalogenation. Kleucker, however, pointed out that the reaction may involve consecutive carbanion formation, alkylation, and dehydrohalogenation.

$$\mathrm{ArCH_2Cl} \xrightarrow{\mathrm{EtO^-}} \mathrm{Ar\overset{\ominus}{C}HCl}$$

$$\downarrow \mathrm{ArCH_2Cl}$$

$$\mathrm{ArCH{=}CHAr} \xleftarrow{-\mathrm{HCl}} \mathrm{Ar\underset{\displaystyle Cl}{\underset{|}{C}}HCH_2Ar}$$

Similarly for the formation of *p,p′*-dinitrostilbene oxide from *p*-nitrobenzaldehyde and *p*-nitrobenzyl chloride in the presence of base either a methylene mechanism or a condensation-dehydrohalogenation mechanism may be written.

$$\mathrm{ArCH_2Cl} \xrightarrow{\mathrm{EtO^-}} \mathrm{Ar\overset{\ominus}{C}HCl} \xrightarrow{-\mathrm{Cl^-}} \mathrm{Ar{-}\ddot{C}{-}H}$$

$$\mathrm{Ar\overset{\ominus}{C}HCl} \xrightarrow{\mathrm{ArCHO}} \mathrm{ArCH(Cl){-}CH(O^{\ominus})Ar} \xrightarrow{-\mathrm{Cl^-}} \mathrm{ArCH{-}CHAr\ (epoxide,\ bridging\ O)}$$

$$\mathrm{Ar{-}\ddot{C}{-}H} \xrightarrow{\mathrm{ArCHO}} \mathrm{ArCH{-}CHAr\ (epoxide,\ bridging\ O)}$$

Hanna, Iskander, and Riad carried out a kinetic investigation to determine whether these reactions of *p*-nitrobenzyl halides with base involve the formation of a methylene or not (144). The reaction of *p*-nitrobenzyl chloride with sodium hydroxide in aqueous dioxane and in aqueous acetone (to give *p,p'*-dinitrostilbene "almost quantitatively") was reported to be first-order in base and first-order in halide. Assuming that no more than a very small fraction of the material is ever present in the form of the various intermediates at any given time, this is the kinetic order that would be expected for the methylene mechanism if either the formation of the carbanion or its decomposition to methylene is rate-controlling. The alkylation-dehydrohalogenation mechanism for the formation of dinitrostilbene would also lead to these kinetics if carbanion formation were the rate-controlling step of the reaction. This cannot be the case, however, because the unreacted *p*-nitrobenzyl chloride recovered from treatment with base in dioxane-deuterium oxide contained about 0.42 deuterium atom per molecule after a time during which about 38 per cent of the halide should have been transformed to dinitrostilbene. This shows that the first step of the reaction is reversible, with the carbanion being protonated at a rate comparable to its rate of further reaction. In this case the reaction rate by the alkylation-dehydrohalogenation mechanism should show a greater-than-first-order dependence on the concentration of *p*-nitrobenzyl chloride, and thus this mechanism may be ruled out on the basis of Hanna, Iskander, and Riad's evidence.

By contrast, the rate of formation of *p,p'*-dinitrostilbene oxide from *p*-nitrobenzaldehyde, *p*-nitrobenzyl chloride, and base was found to increase with increasing aldehyde concentration, showing that the rate-controlling step of the reaction is not methylene formation but probably combination of the aldehyde with a carbanion derived from *p*-nitrobenzyl chloride.

Although the report by Hanna, Iskander, and Riad is in agreement with the hypothesis that a methylene intermediate is formed in the reaction, it gives no direct evidence concerning the mechanism by which the methylene is transformed to *p,p'*-dinitrostilbene. This point will be discussed in relation to a closely related study of the reaction of *p*-nitrobenzyldimethylsulfonium ions with sodium hydroxide that was carried out by Swain and Thornton (145). These workers studied the reaction in dilute aqueous solution (where the total yield of *cis*- and *trans-p,p'*-dinitrostilbene is 99 per cent) in the presence of 0.3 *M* sodium perchlorate to minimize the effect of changing ionic strength. The organic reactant was found to undergo base-catalyzed deuterium exchange much more rapidly than it formed dinitrostilbene, showing that carbanion formation cannot be rate-controlling. The basic reaction solution was slightly orange, presumably because of the presence of small amounts of the carbanion.

$$(CH_3)_2\overset{\oplus}{S}-\overset{\ominus}{C}H-C_6H_4-NO_2 \leftrightarrow (CH_3)_2S=CH-C_6H_4-NO_2 \leftrightarrow (CH_3)_2\overset{\oplus}{S}-CH=C_6H_4=NO_2^{\ominus} \leftrightarrow \text{etc.}$$

When a pellet of potassium hydroxide was added to the solution a deep crimson color surrounded it as it dissolved, showing that the light orange color stemmed from the conversion of only a small fraction of the sulfonium ion to its carbanion.

Since it therefore appears that the steady state assumption can be made for the carbanion, the observation that the reaction is first-order in base and first-order in sulfonium salt rules out an alkylation-elimination mechanism for the reaction, but is in agreement with a mechanism involving rate-controlling loss of dimethyl sulfide from the intermediate carbanion to give *p*-nitrophenylmethylene. This methylene may either dimerize or add to a carbanion and then undergo an elimination reaction.

$$\text{Ar—C—H} \xrightarrow{\text{Ar}-\text{C}-\text{H}} \text{ArCH=CHAr}$$

$$\text{Ar—C—H} \xrightarrow{\text{Ar}\overset{\ominus}{\text{C}}\text{H}\overset{\oplus}{\text{S}}\text{Me}_2} \text{Ar—}\overset{\ominus}{\text{C}}\text{H—}\underset{\oplus\text{SMe}_2}{\text{CH}}\text{—Ar} \xrightarrow{-\text{Me}_2\text{S}} \text{ArCH=CHAr}$$

On the basis of the sulfur kinetic isotope effect, which is less than half as large as that observed in the solvolysis of *t*-butyldimethylsulfonium ions, Swain and Thornton argue that the methylene does not dimerize but adds to carbanions. It is not clear, however, that the kinetic isotope effect in the present case can be predicted with the reliability required by this argument. In either case the 99 per cent yield of dinitrostilbene obtained from 0.1 *M* sulfonium salt and 0.2 *M* sodium hydroxide in aqueous solution (55 *M* water) shows that if a methylene is indeed formed in the reaction it is a very highly selective species since it appears to react with carbanions that are present at a concentration probably no larger than 0.01 *M* or with a methylene whose concentration is probably much smaller, rather than with the much more abundant species hydroxide ion and water. In reaction with the intermediate methylene another methylene molecule has the advantage of leading directly to a stable molecule, but the carbanion probably has the advantage of being present at a higher concentration.

Reactions of Benzyl and Allyl Halides with Alkali Amides. Kharasch and Sternfeld discovered that allyl chloride reacts with sodamide in liquid ammonia to give 1,3,5-hexatriene in yields up to 30 per cent. It was subsequently shown that this reaction may be applied to β-methylallyl chloride and that in both cases, when the allylic chloride was present in excess, a chlorohexadiene could be isolated. Although the structures of the chlorohexadienes were not established this observation suggested that the hexatrienes are probably not formed by the dimerization of methylenes but by the alkylation-dehydrohalogenation mechanism, as follows:

$$CH_2{=}CH{-}CH_2Cl \xrightarrow{NH_2^-} CH_2{=}CH{-}\overset{\ominus}{C}HCl$$

$$\downarrow CH_2{=}CH{-}CH_2Cl$$

$$CH_2{=}CH{-}CH{=}CH{-}CH{=}CH_2 \xleftarrow{-HCl} CH_2{=}CH{-}\underset{\displaystyle Cl}{\underset{|}{C}H}{-}CH_2{-}CH{=}CH_2$$

Analogous treatment of α-phenylethyl chloride yields 2-chloro-2,3-diphenylbutane and then 2,3-diphenyl-2-butene (dimethylstilbene), presumably by an analogous mechanism. Stilbene is formed in 100 per cent yield from benzyl chloride by the action of excess sodamide (but not potassium hydroxide, sodium ethoxide, nor sodium formamide, all of which give benzylamine) in liquid ammonia (but not in ether, where benzylamine was obtained, nor in ligroin, where no reaction occurred) (146). Hauser and coworkers found that in this case, too, the intermediate chloride (1,2-diphenylethyl chloride) can be isolated in as much as 75 per cent yield if an excess of sodamide is avoided (147).

Thus the *p*-nitro substituent appears to have a profound effect on the transformation of benzyl halides to stilbenes by the action of base. As described in the previous section there is kinetic evidence that methylenes are involved in the reactions of the *p*-nitro compound, but, as described in this section, the isolation of intermediate diphenylethyl chlorides provides evidence that the unsubstituted compounds do not react via methylenes. It might be suggested that the diphenylethyl chlorides are indeed formed from methylenes, via combination with carbanions and subsequent protonation, as follows:

$$C_6H_5-C-CH_3 + C_6H_5-\overset{\ominus}{\underset{\displaystyle Cl}{C}}-CH_3 \rightarrow \underset{\text{I}}{C_6H_5-\overset{\ominus}{\underset{\displaystyle CH_3}{\overline{C}}}-\overset{\displaystyle CH_3}{\underset{\displaystyle Cl}{C}}-C_6H_5}$$

$$\downarrow NH_3$$

$$C_6H_5-\underset{\displaystyle CH_3}{CH}-\overset{\displaystyle CH_3}{\underset{\displaystyle Cl}{C}}-C_6H_5$$

This mechanism has several weak points, however. First, the intermediate carbanion I might well be expected to lose a chloride ion to give dimethylstilbene faster than it would be protonated by such a weak acid as ammonia. Perhaps this objection could be met by the hypothesis that carbanion I (or at least a certain fraction of carbanion I) has a conformation that is unfavorable for olefin formation and the carbanion is protonated faster than it rotates into a favorable conformation. An alternate hypothesis is that the carbanion I is never formed but that the intermediate methylene undergoes a concerted attack by carbanion and solvent to yield 2-chloro-2,3-diphenylbutane directly in one step. Another objection to any mechanism involving the reaction of α-phenylethylidene ($C_6H_5-C-CH_3$) with an α-phenyl-α-chloroethyl anion is the fact that racemic products would be expected even if optically active starting materials are used. Hauser and coworkers found that optically active α-phenylethyl chloride reacts with lithium amide in liquid ammonia to give optically active 2-chloro-2,3-diphenylbutane (148). In view of the known ability of organolithium compounds to maintain their configuration at the carbon atom to which the lithium is attached, it would be interesting to learn whether an optically active product is obtained when potassium amide is used. Nevertheless, the observations of Hauser and coworkers provide evidence that the reaction does not involve the intermediate formation of a methylene.

At first thought it may seem that although the *p*-nitro group should have a profound effect on the ease of carbanion formation it may not *thereby* bring about a change in the reaction mechanism since each of the two mechanisms involves an intermediate carbanion. However it is possible that there is also a solvent effect and that the nitro substituent promotes the methylene mechanism by making possible carbanion formation in a hydroxylic solvent. Another possibility, which should certainly be investigated, is that radical-anions, which are known to be formed readily from aromatic nitro compounds, are intermediates in the reaction of *p*-nitrobenzyl compounds with strong bases.

The Reactions of Benzyl Halides and Ethers with Organometallic Compounds. Closs and Closs found that the reaction of benzyl chloride with *n*-butyllithium in ether and cyclohexene leads to the formation of 14 per cent 7-phenylbicyclo-[4.1.0]heptane, among other products (149).

$$C_6H_5CH_2Cl + C_6H_{10} \xrightarrow{n\text{-}C_4H_9Li} C_6H_{10}{>}CH{-}C_6H_5$$

This observation may be explained on the basis of the initial α-dehydrochlorination of benzyl chloride to give phenylmethylene, which then adds to the olefinic double bond; but, in view of the fact that the formation of cyclopropane derivatives is not in itself adequate evidence for the intermediacy of methylenes and the fact that the reaction of benzyl chloride with sodamide may not give phenylmethylene, it is believed that the elucidation of the mechanism of the reaction of benzyl chloride with butyllithium requires further study. The same may be said of the reaction of benzyl phenyl ether with butyllithium, found by Schöllkopf and Eisert to yield a small amount of 1,1-dimethyl-2-phenylcyclopropane when carried out in the presence of isobutylene or a little benzylidenetriphenylphosphine when carried out in the presence of triphenylphosphine (150).

$$C_6H_5CH_2OC_6H_5 + n\text{-BuLi} \xrightarrow{(C_6H_5)_3P} (C_6H_5)_3P{=}CHC_6H_5$$

α-Elimination Reactions of 9-Substituted Fluorenes. $\Delta^{9,9'}$-bifluorene was obtained from 9-bromofluorene and potassium hydroxide in methanol-acetone by Thiele and Wanscheidt, from 9-fluorenyltrimethylammonium hydroxide by Ingold and Jessop, and from 9-chlorofluorene and potassium amide by Hauser and coworkers (147).

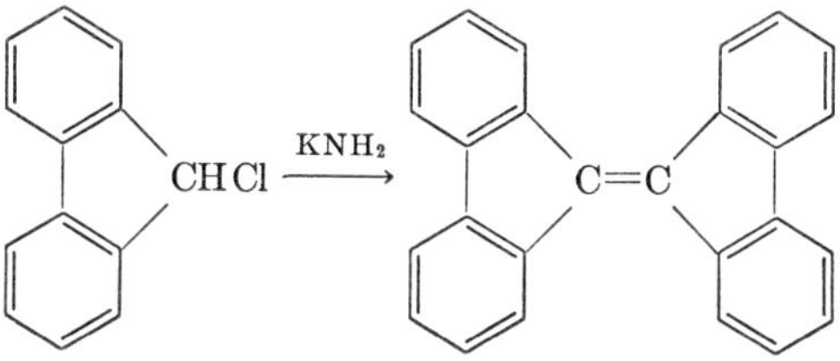

$$\text{(fluorenyl)CHCl} \xrightarrow{KNH_2} \text{(fluorenylidene)C{=}C(fluorenylidene)}$$

Bethell's observations that the formation of $\Delta^{9,9'}$-bifluorene from 9-bromofluorene and potassium *t*-butoxide in *t*-butyl alcohol is first-order in base and second-order in organic halide, and that the base-catalyzed deuterium exchange of 9-deuterio-9-bromofluorene is rapid compared to the formation of bifluorene, show that the alkylation-elimination mechanism is very probably operative (150a).

As shown in the following reaction scheme, a methylene mechanism with 9-fluorenylidene as an intermediate has been suggested for Franzen's observation that the ylid from the

9-fluorenyltrimethylammonium ion reacts with benzyldimethylamine to give 9-benzyl-9-fluorenyldimethylamine (151.)

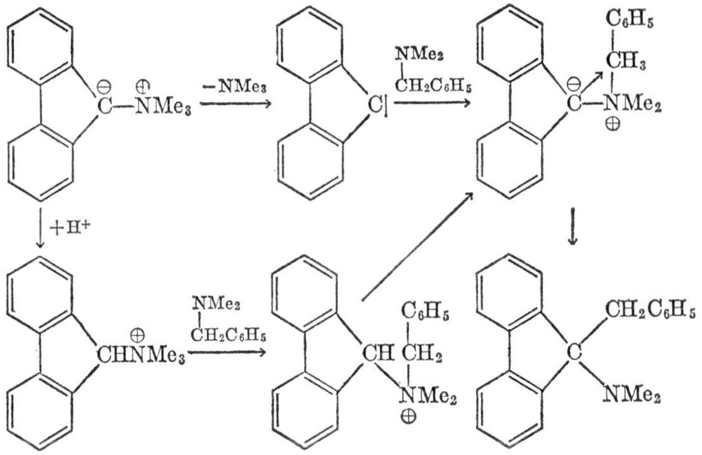

However, as also shown in the reaction scheme, the amine exchange could have occurred by some other reaction mechanism. Perhaps some active hydrogen compound, present as an impurity in the reaction mixture, protonated the initial ylid to give the 9-fluorenyltrimethylammonium ion which then reacted with benzyldimethylamine.

MISCELLANEOUS α-ELIMINATION REACTIONS BROUGHT ABOUT BY STRONG BASES

Reactions of Alkyl Bromides with Alkali-Metal Amides. Hauser and coworkers studied the dehydrohalogenation of several deuteriated saturated aliphatic bromides (152). The dehydrohalogenation of a sample of *n*-octyl bromide, deuteriated in the β-position so that it contained an average of 1.86 ± 0.02 deuterium atoms per molecule, was found to yield 1-octene containing 1.05 ± 0.02 deuterium atoms per molecule. Various assumptions concerning the amount of undeuteriated and monodeuteriated halide in the reactant and the magnitude of the deuterium kinetic isotope effect (k_H/k_D assumed to be between 1 and 10) lead to the conclusion that pure β-dehydrobromination should have given olefin containing 0.75 ± 0.18 atom of deuterium per molecule. The fact that the product contained significantly more deuterium than this can be explained on the basis of a considerable amount of the reaction consisting of α-dehydrobromination and rearrangement. This process is shown below as involving an intermediate carbanion and an intermediate methylene. There is no direct evidence for either of these intermediates and the reaction may be partially or completely concerted. However, the plausibility of a methylene intermediate in the reaction is considerably increased by the evidence for methylenes as intermediates in similar reactions.

$$n\text{-}C_6H_{13}CD_2CH_2Br \xrightarrow{NH_2^-} n\text{-}C_6H_{13}CD_2\overset{\ominus}{C}HBr$$

$$\downarrow -Br^-$$

$$n\text{-}C_6H_{13}CD{=}CHD \leftarrow n\text{-}C_6H_{13}CD_2\text{—}C\text{—}H$$

In view of the subsequent investigations to be described in the next section, it seems possible that the excess deuterium may have been present in some n-pentylcyclopropane present as an impurity in the reaction product.

$$n\text{-}C_5H_{11}CH_2CD_2CH_2Br \xrightarrow{NH_2^-} n\text{-}C_5H_{11}CH\langle\begin{matrix}CD_2\\ |\\ CH_2\end{matrix}$$

Reactions of Alkyl Chlorides with Strong Bases. Whitmore and coworkers observed that the reaction of neopentyl chloride with sodium yields largely 1,1-dimethylcyclopropane, as does the reaction with n-propylsodium.

$$CH_3\text{—}\underset{CH_3}{\overset{CH_3}{\underset{|}{\overset{|}{C}}}}\text{—}CH_2Cl + n\text{-}C_3H_7Na \rightarrow CH_3\text{—}\overset{CH_3}{\overset{|}{C}}\text{——}CH_2 + C_3H_8 + NaCl$$
$$\diagdown\ \diagup$$
$$CH_2$$

These reactions and the analogous reactions of neohexyl chloride were formulated as γ-eliminations. Kirmse and Doering, however, showed that the formation of cyclopropane derivatives from alkyl chlorides and sodium (or organosodium compounds) occurs even with alkyl chlorides that have β-hydrogen substituents, and that the reactions involve α- rather than γ-elimination (153). The reaction of sodium in cyclohexane with *n*-propyl chloride gives cyclopropane as 4 per cent of the C_nH_{2n} fraction, *n*-butyl chloride gives 7 per cent methylcyclopropane, and isobutyl chloride gives 35 per cent methylcyclopropane. Although potassium works as well as sodium, lithium gives no methylcyclopropane. The reaction of isobutyl *bromide* and sodium gives only 3 per cent methylcyclopropane. The reaction of sodium with 1,1-dideuterioisobutyl chloride was found to give monodeuteriomethylcyclopropane containing less than one per cent dideuteriated material. Thus it seems clear that the methylcyclopropane is not formed by a γ-elimination but by an α-elimination and cyclization as shown in the following reaction mechanism.

$$RCl + 2Na \rightarrow NaCl + RNa$$

$$CH_3\text{—}CH(CH_3)\text{—}CD_2Cl + RNa \rightarrow CH_3\text{—}CH(CH_3)\text{—}CD(Na)Cl + RH$$

$$CH_3\text{—}CH(CH_3)\text{—}CD(Na)Cl \xrightarrow{-NaCl} CH_3\text{—}CH(CH_3)\text{—}\ddot{C}\text{—}D$$

$$CH_3\text{—}CH(CH_3)\text{—}C\text{—}D \rightarrow CH_3\text{—}\underset{\diagdown\ CH_2\ \diagup}{CH\text{——}CHD}$$

There appears to be no direct evidence that the α-sodio compound shown is really a reaction intermediate and therefore it is possible that the α-elimination is concerted. Among the evidence that the methylene shown is an intermediate is the fact that about the same ratio of alkene to cyclopropane is obtained in the reactions of *n*-propyl, *n*-butyl, isobutyl, and neopentyl chlorides with sodium as in the corresponding alkaline decompositions of aldehyde tosylhydrazones, which are also believed to involve the formation of intermediate methylenes (cf. Chapter 7).

Phenylsodium and sodamide are weaker bases than alkylsodium compounds but they, too, give largely α-elimination upon reaction with neopentyl chloride. Friedman and Berger found that both of these bases give 1,1-dimethylcyclopropane upon reaction with neopentyl chloride (154). Phenylsodium and 1,1-dideuterioneopentyl chloride yield monodeuteriobenzene, monodeuterio-1,1-dimethylcyclopropane, and a little monodeuterio-2-methyl-2-butene.

$$(CH_3)_3CCD_2Cl \xrightarrow{C_6H_5Na} (CH_3)_3C\text{—}\ddot{C}\text{—}D$$

$$\swarrow \qquad \downarrow$$

$$(CH_3)_2C{=}CDCH_3 \qquad (CH_3)_2C\underset{CH_2}{\diagdown\!\!\!\!\text{——}\!\!\!\!\diagup}CHD$$

The olefin and cyclopropane apparently arise from an intermediate methylene by a shift of a methyl group and an internal insertion reaction, respectively. The olefin:cyclopropane ratios, ranging from 11.7:1 to 31:1, may be compared with the 10:1 ratio obtained in the alkaline decomposition of the tosylhydrazone of 2,2-dimethylpropanal. The differences in ratios obtained may be no larger than the experimental uncertainty, but the 31:1 ratio was obtained in the same solvent as the 10:1 ratio.

Reaction of Cycloalkene Oxides with Lithium Diethylamide. Cope and coworkers found that the reaction of *cis*-cyclodecene oxide with lithium diethylamide yields *cis*-*cis*-1-decalol (III) and smaller amounts of *endo*-*cis*-bicyclo-[5.3.0]decan-2-ol (IV) and 2-cyclodecen-1-ol (V).

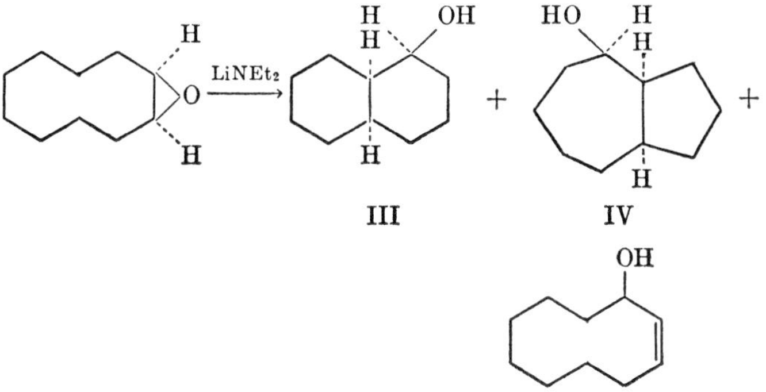

H H OH HO H
H H
LiNEt₂
O
+ +
H H
H
III IV
OH

V

The reaction of *cis*-cycloöctene oxide is similar, yielding largely *endo-cis*-bicyclo[3.3.0]octan-2-ol.

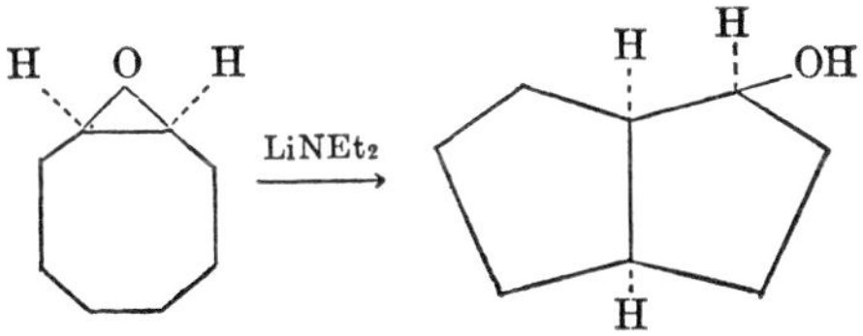

H
H O H H OH
LiNEt₂

H

These reactions, consisting of the transformation of $C_{10}H_{18}O$ to $C_{10}H_{17}OH$ and of $C_8H_{14}O$ to $C_8H_{13}OH$, obviously involve the removal of a carbon-bound hydrogen atom. If the reactions are 1,6- and 1,5-eliminations the base must remove hydrogen from a carbon atom across the ring from the epoxide group. On the other hand if the reactions are 1,1-eliminations to give methylenes, which subsequently (or simultaneously) perform internal insertion reactions, the hydrogen atom removed will be one attached to the same carbon atom as the epoxy oxygen atom. Cope and coworkers used this fact to distinguish between the two possible types of mechanisms (155). The reaction of 1,2-dideuterio-*cis*-cyclodecene oxide with lithium diethylamide was found to give the bicyclic alcohols III and IV containing only one deuterium atom per molecule. The reaction is thus a 1,1-elimination, which may proceed by the stepwise mechanism shown below.

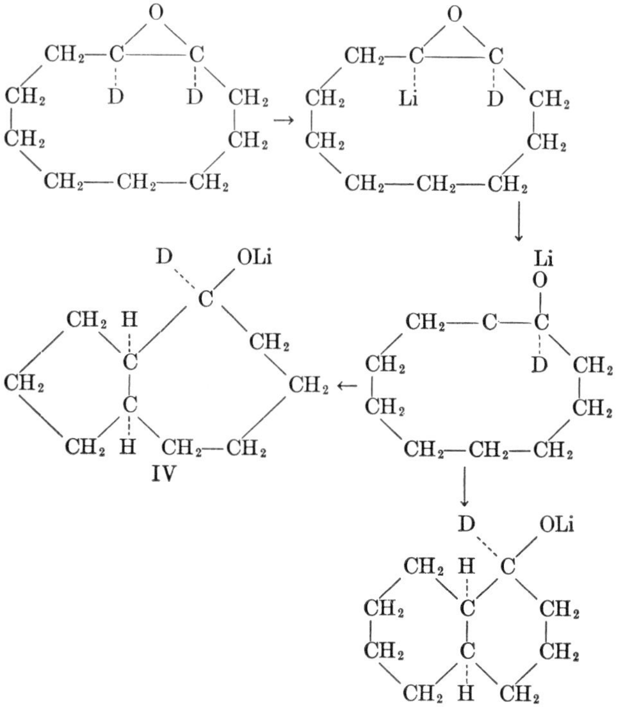

IV

III

Concerted reaction mechanisms may also be envisaged in which the intermediate organolithium compound, or the methylene, or both, are not real reaction intermediates. When *trans*-cyclodecene oxide is used instead of the *cis* isomer, *cis-trans*-1-decalol (VI) is obtained as the major product, instead of *cis-cis*-1-decalol.

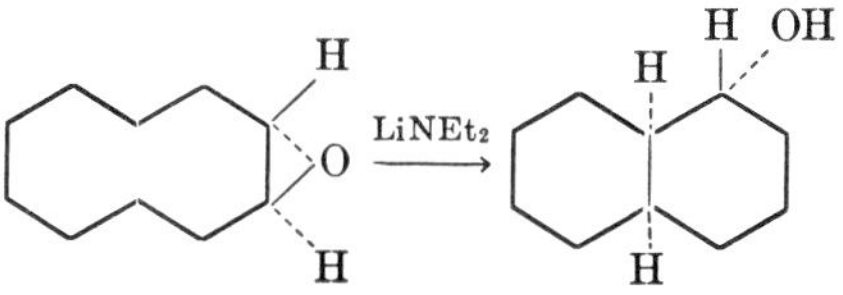

$\xrightarrow{LiNEt_2}$

VI

This is the result expected from a concerted reaction mechanism in which there is no methylene formed as a true intermediate, but it may also be explained by a mechanism involving the formation of an intermediate methylene that undergoes transannular insertion faster than it undergoes conformational change.

The results obtained in the cycloöctane series are analogous. The reaction of 5,6-dideuterio-*cis*-cycloöctene oxide with lithium diethylamide yields a dideuteriated *endo-cis*-bicyclo-[3.3.0]octan-2-ol.

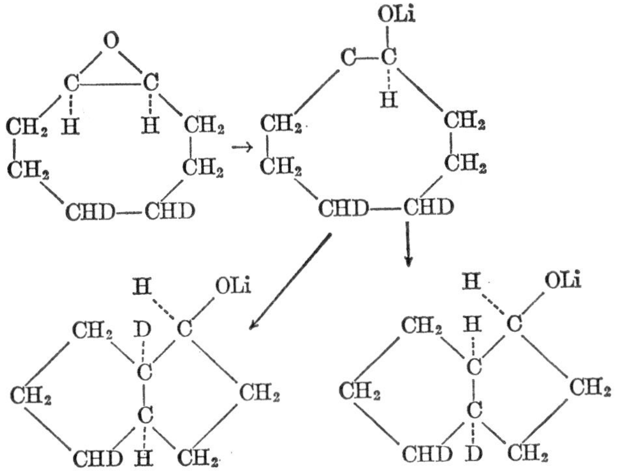

From *trans*-cycloöctene oxide, *exo-cis*-bicyclo[3.3.0]octan-2-ol was obtained.

7

Reactions of Substituted Methylenes Formed from Diazo Compounds

INTRODUCTION

In Chapter 2 convincing evidence that the decomposition of diazomethane can lead to free methylene was described. It is highly probable that substituted methylenes may be formed by the decomposition of derivatives of diazomethane but in no case is the present evidence quite as strong as with diazomethane itself. Even with diazomethane many of the reactions for which a methylene intermediate could be written proceed by another mechanism. With many derivatives of diazomethane still more alternative reaction mechanisms are possible. The formation of methylenes from diazo compounds has been the subject of much conjecture but relatively little careful mechanistic (especially kinetic) study. For this reason the reader should view most of the reaction mechanisms written in the current chapter as plausible but not firmly established, even though the probable validity of the various mechanisms is commented on only occasionally.

The chemistry of diazo compounds, including their transformations to methylenes, has been the subject of a penetrating review by Huisgen (5). The most significant new method of generating diazo compounds beyond those described by Huisgen was discovered by Bamford and Stevens, investigated further by Powell and Whiting (156), and applied to the generation of methylene derivatives by Friedman and Shechter (157). This method involves the treatment of *p*-toluenesulfonylhydrazones with base at 100–200°. Apparently the anion of the hydrazone loses a *p*-toluenesulfinate ion to give the diazo compound, which may then react further. In hydroxylic solvents varying amounts of products that would be expected from intermediate carbonium ions are formed, probably by protonation of the intermediate diazo compound.

$$R_2C{=}N{-}\overset{\ominus}{N}{-}SO_2C_7H_7 \rightarrow R_2C{=}N{=}N + C_7H_7SO_2^-$$

$$R_2C{=}N{=}N \xrightarrow{-N_2} R{-}C{-}R$$

$$R_2C{=}N{=}N \xrightarrow{ROH} R_2CH{-}N{\equiv}\overset{\oplus}{N} \xrightarrow{-N_2} R_2\overset{\oplus}{CH}$$

In less acidic solvents increasing amounts of products whose formation can be explained on the basis of a methylene intermediate are observed.

Diazoketones comprise one of the most important classes of diazo compounds. Their use in synthetic organic chemistry has been reviewed critically by Weygand and Bestmann (158).

Methylenes may also be generated from derivatives of cyclodiazomethane, which may be prepared, in at least some cases, by methods analogous to those used for cyclodiazomethane itself.

INSERTION OF METHYLENES INTO CARBON-HYDROGEN BONDS

Internal Carbon-Hydrogen Insertions in the Camphor Series. Since intramolecular reactions yielding cyclic products often occur much more rapidly than their intermolecular counterparts it is not surprising that certain types of methylenes appear to undergo internal insertion reactions even when no insertion reactions with external reagents can be detected. Bredt and Holz showed that the previously observed product of the decomposition of diazocamphor at 140°, dehydrocamphor, is a tricyclic compound containing a three-membered ring.

```
                                                              Me
                                                              |
        Me                    Me                              C
        |                     |                             / | \
CH2 --- C --- CO       CH2 --- C --- CO         CH2         |   CO
|       |      |  -N2  |       |      |          |       CMe2    |
|      CMe2    |  -->  |      CMe2    |   -->    |         |     |
|       |      |       |       |      |          |        CH     |
CH2 --- CH --- CN2     CH2 --- CH --- C          |      /    \   |
                                                 CH ----------- CH
```

Analogously Meerwein and van Emster showed that the mercuric-oxide oxidation of camphor hydrazone leads to tricyclene, presumably via an intermediate diazo compound and a methylene.

```
           Me                               Me
           |                                |
           C                                C
         / | \                            / | \
     CH2   |   C=NNH2       HgO       CH ---|--- CH
      |   CMe2  |           -->        |   CMe2   |
     CH2   |   CH2                    CH2   |    CH2
        \  |  /                          \  |  /
          CH                               CH
   Camphor hydrazone                   Tricyclene
```

Among the evidence for a methylene intermediate in this case is the fact that the corresponding carbonium ion is known to give other products. Powell and Whiting found that tricyclene may also be prepared in high yield by the decomposition of the sodium salt of the *p*-toluenesulfonylhydrazone of camphor in a weakly acidic solvent, such as acetamide, or by the decomposition of pre-formed diazocamphane (the yield was poorer in this case, perhaps partly because the more acidic solvent 2-ethoxyethanol was used) (156).

Other Internal Carbon-Hydrogen Insertion Reactions. Friedman and Shechter found that the decomposition of the *p*-toluenesulfonylhydrazones of certain cyclic ketones results in internal insertion with the formation of bicyclic compounds (157). From the cyclopentanone and cyclohexanone derivatives quantitative yields of cyclopentene and cyclohexene, respectively, were obtained. The latter result is in contrast to the large extent of internal insertion observed with camphor and norcamphor and is to be attributed to the fact that the six-membered ring of the bicyclic compounds is in a boat conformation whereas that of the monocyclic compound is in a chair conformation. The derivatives of cycloheptanone, cycloöctanone, cyclononanone, and cyclodecanone gave bicyclic products via 1,3-, 1,5-, and 1,6-insertion processes in yields of 18 per cent, 55 per cent, 76 per cent, and 80 per cent, respectively, with the remainder of the products being cycloalkenes. Thus, from cyclodecanone *p*-toluenesulfonylhydrazone, 62 per cent *cis*-bicyclo[5.3.0]decane, 18 per cent *cis*-decalin, 14 per cent *cis*-cyclodecene, and 6 per cent *trans*-cyclodecene were formed.

$HNSO_2C_7H_7$
N
base →
H H
+ +
H H
+

Frey and Stevens found that an intramolecular insertion reaction of a methylene may be affected by the amount of excess energy the methylene has, just as intermolecular reactions of methylenes are. The thermal decomposition of 2-diazobutane at 160° was found to give 0.5 per cent methylcyclopropane, 3.6 per cent 1-butene, 29.5 per cent *cis*-2-butene, and 66.5 per cent *trans*-2-butene, in good agreement with the results obtained by Friedman and Shechter in the basic decomposition of the *p*-toluenesulfonylhydrazone of methyl ethyl ketone (157).

$$CH_3CH_2\text{—}\overset{\overset{\displaystyle N_2}{\|}}{C}\text{—}CH_3 \rightarrow CH_3CH_2\text{—}\ddot{C}\text{—}CH_3 \rightarrow \text{methylcyclopropane } (CH_3\text{—}CH\text{ in ring with } CH_2\text{—}CH_2)$$

$$\swarrow \quad \downarrow \quad \searrow$$

$$\underset{trans}{(CH_3)(H)C{=}C(H)(CH_3)} \qquad \underset{cis}{(H)(CH_3)C{=}C(H)(CH_3)} \qquad CH_3CH_2CH{=}CH_2$$

The photolysis of 2-diazobutane, using light of wave length 3130 Å, yields 2.4 per cent methylcyclopropane, 23.4 per cent 1-butene, 35.6 per cent *cis*-2-butene, and 38.6 per cent *trans*-2-butene when the pressure is high enough (>50 mm.) to avoid secondary isomerization of initially formed "hot" products (159). In both cases the yields of the various products increase with the increasing stability of the products. In the photolysis reaction, however, where the intermediate methylene has far more excess energy, there is a much smaller difference between the yields of the various products, that is, the reaction proceeds in a more nearly random fashion.

Intermolecular Carbon-Hydrogen Insertion Reactions. In general, substituted methylenes display a smaller tendency to give intermolecular insertion reactions than does methylene itself. In many cases it is difficult to tell to what extent this is due to decreased reactivity in the insertion reaction and to what extent it merely reflects the availability of more attractive reaction paths such as rearrangement, internal insertion, etc. The first intramolecular insertion of a substituted methylene into a carbon-hydrogen bond appears to have been observed by Buchner and Schulze, who found that the decomposition of methyl or ethyl diazoacetate in boiling *p*-xylene leads to an ester of *p*-methylhydrocinnamic acid, in addition to the product of ring expanson.

$$CH_3-C_6H_4-CH_3 + N_2CHCO_2R \xrightarrow{-N_2} CH_3-C_6H_4-CH_2CH_2CO_2R$$

Smith and Tawney found that with durene carbon-hydrogen insertion becomes the principal reaction path, 2,4,5-trimethylhydrocinnamic acid esters being isolated in about 30 per cent yield.

Doering and Knox discovered that carbethoxymethylene is capable of attack on saturated hydrocarbons (1). With cyclopentane and cyclohexane the ethyl esters of the corresponding cycloalkylacetic acids are obtained.

$$C_6H_{12} + N_2CHCO_2Et \xrightarrow{h\nu} C_6H_{11}-CH_2CO_2Et$$

The reaction with *n*-pentane yielded a mixture of ethyl esters of $C_6H_{13}CO_2H$ acids.

Apparently α-carbalkoxy substituents decrease the reactivity of methylene in carbon-hydrogen insertion reactions with saturated hydrocarbons. When 2,3-dimethylbutane is used as the hydrocarbon reactant the ratio of the reactivity of the tertiary hydrogens to that of the primary hydrogens increased from 1.2 for attack by methylene to 2.9 for attack by carbomethoxymethylene to 12.5 for attack by bis(carbethoxy)methylene (160). These observations may be explained simply by the hypothesis that methylene is so reactive that a carbon-hydrogen insertion reaction occurs at almost every collision of methylene with a hydrocarbon molecule; when the stabilizing carbalkoxy substituents are present the decreased reactivity of the methylene permits the differences in reactivity of the various carbon-hydrogen bonds to be manifest. As Doering and Knox pointed out, however, a number of other factors may be important. The reacting species in the present case may be, to a large extent, methylene as it is originally formed in the photolysis reaction. This species is much different from thermally equilibrated methylene, which undoubtedly has less energy, apparently has a different multiplicity, may react by a different mechanism, and certainly shows greater selectivity (cf. Chapter 2). Carbalkoxymethylenes may have less excess energy and may give insertion reactions by a different mechanism from initially formed methylene. Furthermore, the ability of carbalkoxy substituents to stabilize a negative charge permits the stabilization of the insertion transition state by the contribution of zwitterionic structures that must be much more stable for tertiary than for primary R groups.

```
                            ⊕
  R———H                   R     H
                            ⊖  /
      C          ↔           C         ↔ etc.
     / \                    / \
    H   C=O                H   C=O
        |                      |
        OMe                    OMe
```

A phenyl substituent also makes methylene more selective in its carbon-hydrogen insertion reactions, and in this case the selectivity is not so readily attributed to the electron-withdrawing power of the substituent. Gutsche, Bachman, and Coffey found that the photolysis of phenyldiazomethane in *n*-pentane gives about six times as much insertion of the benzylidene group into the six secondary carbon-hydrogen bonds as into the six primary carbon-hydrogen bonds (161).

Relatively little is known about which substituted methylenes give carbon-hydrogen insertion by the direct one-step mechanism and which ones react by the two-step mechanism involving the formation and coupling of two radicals (cf. Chapter 2). Parham and Hasek obtained evidence that diphenylmethylene abstracts hydrogen atoms from saturated hydrocarbons. They found that the decomposition of diphenyldiazomethane in hot petroleum ether yields tetraphenylethane (in addition to benzophenone azine) (162).

$$(C_6H_5)_2CN_2 \rightarrow C_6H_5\text{—}C\text{—}C_6H_5 + N_2$$

$$C_6H_5\text{—}C\text{—}C_6H_5 + RH \rightarrow C_6H_5\text{—}\dot{C}H\text{—}C_6H_5 + R\cdot$$

$$2C_6H_5\text{—}\dot{C}H\text{—}C_6H_5 \rightarrow (C_6H_5)_2CH\text{—}CH(C_6H_5)_2$$

Kirmse, Horner, and Hoffman found that the yield of tetraphenylethane in the thermal decomposition of diphenyldiazomethane increases with the concentration and reactivity of carbon-hydrogen bonds in the solvent. Only a trace of tetraphenylethane was formed in benzene (aromatic carbon-hydrogen bonds are stronger than aliphatic carbon-hydrogen bonds); in cyclohexane a 29 per cent yield of tetraphenylethane was obtained; toluene, which has fewer but more reactive carbon-hydrogen bonds, gave 35 per cent tetraphenylethane; and with a 20 per cent solution of diphenylmethane in benzene 53 per cent tetraphenylethane was obtained (163). No benzhydrylcyclohexane could be isolated when diphenyldiazomethane was photolyzed in cyclohexane, but when a 10 per cent solution of fluorene in benzene was used as the solvent a 33 per cent yield of 9-benzhydrylfluorene was obtained.

The photolysis of 9-diazofluorene and diazocyclopentadiene in cyclohexane gave rather different results from those obtained with diphenyldiazomethane. With diazofluorene both 9-cyclohexylfluorene and 9,9′-difluorenyl were obtained. Diazocyclopentadiene yielded 57 per cent cyclopentadienylcyclohexane but no dicyclopentadienyl (163).

The explanation of this difference in behavior is not apparent. It might be suggested that the methylenes obtained on the photolysis of diazofluorene and diazocyclopentadiene have a much greater tendency to exist in the singlet form than does diphenylmethylene, and therefore these singlet methylenes give the concerted one-step insertion reaction that appears to be characteristic of unsubstituted singlet methylene (cf. Chapter 2). There is certainly good reason to expect a greater relative stability of the singlet form for the cyclic methylenes. They can assume the configuration (linear at the divalent carbon atom) expected for optimum stability of the triplet form only by introducing an exorbitant amount of strain into their rings. However, Trozzolo, Murray, and Wasserman obtained direct evidence by EPR measurements on an irradiated 9-diazofluorene-containing glass at 77° K that biphenylenemethylene (i.e., 9-fluorenylidene) is a triplet in the ground state (163a).

INSERTION OF METHYLENES INTO CARBON-CARBON SINGLE BONDS

Although many insertions into carbon-hydrogen bonds are known, early attempts to find evidence for insertions into carbon-carbon single bonds were unsuccessful. Doering and Knox, for example, found no evidence for ring expansion in the reaction of carbethoxymethylene with cyclopentane (1). According to Kirmse, however, Doering and Jones subsequently observed a small amount of insertion into a carbon-carbon bond in the reaction of the highly strained hydrocarbon spiropentane with methylene (4). Yates and Danishefsky found that intramolecular insertion into a carbon-carbon single bond may proceed in good yield. The copper-bronze-catalyzed decomposition of 1,5,5-trimethyl-3-azobicyclo[2.2.1]-heptan-2-one (I) gives more than 50 per cent of the tricyclic ketone (II) (164).

```
                                                                  CH3
                                                                   |
                                                                   C
         CH3                         CH3                        /  |  \
          |                           |                        /   |   \
   CH2—C——CO                 CH2—C——CO               CH2   |    C=O
    |      |      |      Cu        |      |      |               |   CH2    |
    |     CH2     |      →         |     CH2     |       →       |    |     |
    |      |      |                |      |      |               |   CH     |
CH3—C——CH—CN2              CH3—C——CH—C                   |  /   \    |
    |                              |                             | /     \   |
   CH3                            CH3                            C———————C
                                                                 |         |
                                                                CH3       CH3
        I                                                            II
```

Although a methylene is written as an intermediate in the reaction scheme above, it is quite possible that no methylene is actually involved; the reactive intermediate may be an organocopper compound instead. The same possibility exists for the following internal insertion into a carbon-carbon single bond reported by Lansbury and Colson (165).

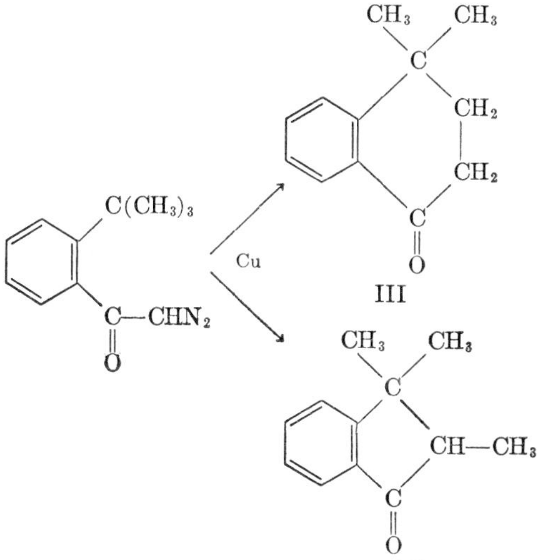

About one per cent of the carbon-carbon insertion product IV (but no III) was obtained when α-diazo-*o*-*t*-butylacetophenone was decomposed in the presence of copper in dimethyl sulfoxide solution but 12 per cent III and no IV was found when the reaction was carried out in benzene solution. It was therefore suggested that the transition state for the carbon-carbon insertion reaction is much more polar than the transition state for carbon-hydrogen insertion.

ATTACK OF METHYLENES ON AROMATIC RINGS

The Attack of Carbalkoxymethylenes on Aromatic Rings. Buchner and coworkers discovered that the thermal decomposition of methyl and ethyl diazoacetate in the presence of an excess of benzene yields an unsaturated ester that can be transformed in various ways to other unsaturated esters and the corresponding acids, which can be reduced to cycloheptanecarboxylic acid. Doering and coworkers later showed that the initial reaction product is an ester of 2,4,6-cycloheptatrienecarboxylic acid (166).

$$C_6H_6 \xrightarrow{N_2CHCO_2R} C_7H_7\text{—}CO_2R$$

Because of the tendencies toward interaction of the ends of the unsaturated system (as shown by Diels-Alder reactions of the type described below), Doering prefers to write the structure with the partial (dotted) bond shown at the right above.

The decomposition of diazoacetates in the presence of aromatic compounds has been of particular importance as a method of entry into the azulene and tropylium series. The reaction may be brought about by light as well as by heat, and although no careful mechanistic study seems to have been made it is probable that intermediate carbalkoxymethylenes (H—C—CO_2R) are involved. Addition to a number of derivatives of benzene has been carried out but in many cases the structure of the immediate product(s) was not determined (it was often possible to transform all the isomeric immediate products into the desired final product).

Alder and coworkers studied the products of reaction of carbomethoxymethylene with several alkylbenzenes and found that the methylene usually avoids reaction at an alkylated ring-carbon atom. The reaction of toluene, for example, gave a mixture of the methyl esters of methylcycloheptatrienecarboxylic acids. The reaction of this mixture (shown below for the isomer produced by attack on the C_2—C_3 bond of toluene) with dimethyl acetylenedicarboxylate followed by thermal "extrusion" of a molecule of a methyl ester of a

Although a methylene is written as an intermediate in the reaction scheme above, it is quite possible that no methylene is actually involved; the reactive intermediate may be an organocopper compound instead. The same possibility exists for the following internal insertion into a carbon-carbon single bond reported by Lansbury and Colson (165).

$C(CH_3)_3$
C—CHN_2
O
Cu
CH_3 CH_3 C CH_2 CH_2 C O
III
CH_3 CH_3 C CH—CH_3 C O
IV

About one per cent of the carbon-carbon insertion product IV (but no III) was obtained when α-diazo-*o*-*t*-butylacetophenone was decomposed in the presence of copper in dimethyl sulfoxide solution but 12 per cent III and no IV was found when the reaction was carried out in benzene solution. It was therefore suggested that the transition state for the carbon-carbon insertion reaction is much more polar than the transition state for carbon-hydrogen insertion.

ATTACK OF METHYLENES ON AROMATIC RINGS

The Attack of Carbalkoxymethylenes on Aromatic Rings. Buchner and coworkers discovered that the thermal decomposition of methyl and ethyl diazoacetate in the presence of an excess of benzene yields an unsaturated ester that can be transformed in various ways to other unsaturated esters and the corresponding acids, which can be reduced to cycloheptanecarboxylic acid. Doering and coworkers later showed that the initial reaction product is an ester of 2,4,6-cycloheptatrienecarboxylic acid (166).

$$\text{C}_6\text{H}_6 \xrightarrow{N_2CHCO_2R} \text{(cycloheptatriene)}\text{—}CO_2R$$

Because of the tendencies toward interaction of the ends of the unsaturated system (as shown by Diels-Alder reactions of the type described below), Doering prefers to write the structure with the partial (dotted) bond shown at the right above.

The decomposition of diazoacetates in the presence of aromatic compounds has been of particular importance as a method of entry into the azulene and tropylium series. The reaction may be brought about by light as well as by heat, and although no careful mechanistic study seems to have been made it is probable that intermediate carbalkoxymethylenes (H—C—CO_2R) are involved. Addition to a number of derivatives of benzene has been carried out but in many cases the structure of the immediate product(s) was not determined (it was often possible to transform all the isomeric immediate products into the desired final product).

Alder and coworkers studied the products of reaction of carbomethoxymethylene with several alkylbenzenes and found that the methylene usually avoids reaction at an alkylated ring-carbon atom. The reaction of toluene, for example, gave a mixture of the methyl esters of methylcycloheptatrienecarboxylic acids. The reaction of this mixture (shown below for the isomer produced by attack on the C_2—C_3 bond of toluene) with dimethyl acetylenedicarboxylate followed by thermal "extrusion" of a molecule of a methyl ester of a cyclopropenecarboxylic acid yielded about 60 per cent dimethyl 4-methylphthalate, 35 per cent dimethyl 3-methylphthalate, and 5 per cent dimethyl phthalate (167).

```
                                                   CH3
                                                   |
                  CH3                              C
                  |                   O     CH  //   \  CH
H—C—CO2CH3 +  [benzene]  →  CH3OCCH                     ||
                                         \  CH          CH
                                             \\ CH /

                                  | CO2CH3—C≡C—CO2CH3
                                  ↓
                                                   CH3
                                                   |
            CH3                                    C       CO2CH3
CH          |      CO2CH3             O      CH  CH  C
|| \                                  ||
||  CHCO2CH3 + [benzene]      ← CH3OCCH
|| /                                         CH  CH  C
CH                 CO2CH3                       CH        CO2CH3
```

Apparently, then, 60 per cent of the attack of carbomethoxymethylene on toluene was at the C_3—C_4 bond, 35 per cent at the C_2—C_3 bond, and 5 per cent at the C_1—C_2 bond.

The Attack of Other Methylenes on Aromatic Rings. Treibs and coworkers found that the decomposition of aryldiazomethanes and of α-diazoketones in the presence of aromatic compounds leads to the formation of compounds with seven-membered rings (168). Gutsche and Johnson showed that aromatic-ring expansion can be carried out as an intramolecular reaction (169). The irradiation of a petroleum-ether solution of 2-(2-phenylethyl)phenyldiazomethane led to the formation of the partially hydrogenated cyclohepta-naphthalene shown, in 9 per cent yield, but the insertion product 2-phenylindane was formed in 30 per cent yield.

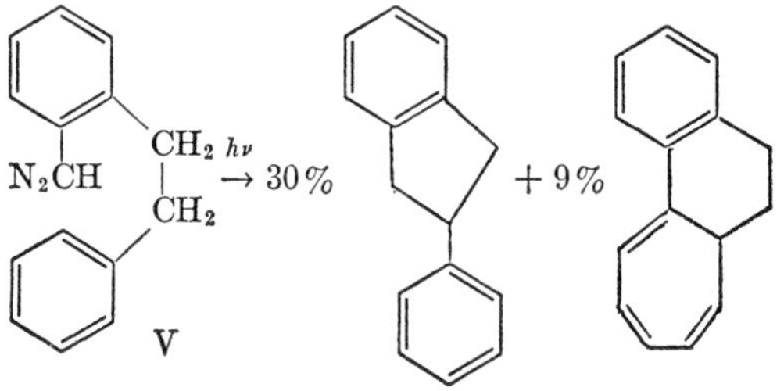

To rationalize the results of this and related experiments Gutsche and coworkers made a study of the inherent relative reactivities of aromatic rings and carbon-hydrogen bonds, of the geometrical factors involved in intramolecular reactions, and of the effect of experimental conditions on the course of the reactions. The presence of a *p*-methoxy substituent in the phenyl group of V was found to have no significant effect on the yield of the product in which the aromatic ring has been expanded (170). When the phenyl group is one carbon atom further away from the diazo group, as in 2-(3-phenylpropyl)-phenyldiazomethane (VI), little, if any, attack on the aromatic ring occurred.

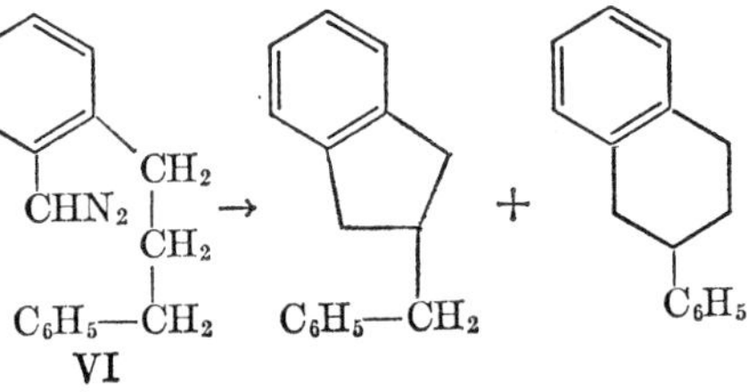

As shown above, two cyclization products, 2-benzylindan and 2-phenyltetralin, were obtained (in a combined yield of 15–25 per cent). The indan was the predominant product under all conditions but the selectivity decreased as the reaction temperature increased. When VI was photolyzed at 15° the indan:decalin ratio was 9:1, at 65° it was 6:1, and the thermal decomposition of VI at 175° gave only twice as much indan as tetralin. If the internal insertion proceeds by the hydrogen abstraction-radical coupling mechanism the preference for indan formation would seem to be due to geometrical factors since the benzyl-type radical that would be an intermediate in tetralin formation should certainly be more stable than the intermediate radical for indan formation. On the other hand no such preference for 1,5-insertion was observed in the photolysis of 2-*n*-butylphenyldiazomethane, where the formation of a seven-, six-, and five-membered ring occurred in the ratio 1:5:6 (161). The small amount of 1,7-insertion may be explained as a geometrical effect or by the lower reactivity of CH_3 groups (relative to CH_2); there certainly appears to be little difference in ease of 1,5- and 1,6-insertion, however.

In order to compare the reactivities of aromatic rings and the carbon-hydrogen bonds of saturated hydrocarbons, Gutsche, Bachman, and Coffey photolyzed phenyldiazomethane in an equimolar mixture of benzene and cyclohexane (161). The ratio of benzylcyclohexane to phenylcycloheptatriene was found to be 1.16:1.00, showing the reactivities to be comparable. The complete absence of diphenylmethane from the reaction product shows how much less reactive the aromatic carbon-hydrogen bond is toward attack by the reactive intermediate benzylidene. Insertion at an aromatic carbon-hydrogen bond by an arylmethylene can be observed, however, as an intramolecular reaction when the geometry is favorable. The photolysis of 2-phenylphenyldiazomethane gives fluorene in high yield.

With at least one methylene with strongly electron-withdrawing substituents, attack on the aromatic ring follows the course to be expected for electrophilic aromatic substitution. This was observed by Weygand and coworkers who photolyzed trifluoroacetyldiazoacetic ester in aromatic hydrocarbons and in chloro- and bromobenzene. Substitution in the *ortho* and *para* positions of the rings was observed (171).

$$CF_3-\overset{\overset{O}{\|}}{C}-\underset{\underset{N_2}{\|}}{C}-CO_2Et \xrightarrow{h\nu} CF_3-\overset{\overset{O}{\|}}{C}-\ddot{C}-CO_2Et \xrightarrow{C_6H_5CH_3}$$

$$p\text{-}CH_3C_6H_4-CH(CO_2Et)-C(=O)-CF_3 \longleftarrow \left[\overset{\oplus}{\text{(CH}_3\text{-cyclohexadienyl)}}(H)-\overset{\ominus}{C}(CO_2Et)-C(=O)-CF_3 \longleftrightarrow \text{etc.} \right]$$

The strongly electron-withdrawing substituents attached to the negatively charged carbon atom in the resonance-stabilized intermediate make the zwitterionic structure particularly stable (relative to a cyclopropane structure) in this case.

ADDITION OF METHYLENES TO MULTIPLE BONDS

The Active Agent in Reactions of Diazo Compounds with Unsaturated Compounds. Many diazo compounds, when heated with various unsaturated compounds, have been found to yield products containing three-membered rings. For such a reaction one may always consider the possibility that the diazo compound decomposes to a methylene, which then adds to the multiple bond. Another possibility, however, is that the diazo compound itself adds to the multiple bond to give an intermediate (a pyrazoline, as shown below, when the multiple bond is a carbon-carbon double bond), which then loses nitrogen to give the three-membered-ring product, e.g.:

```
                                   CO2R
                                    |
CO2R        CO2R                    CH
 |           |                     /  \
 CH          CH              N          CH—CO2R
 ||    +     ||      →       ||         |
 N           CH              N——————————CH
 ||          |                          |
 N           CO2R                       CO2R

                                    | −N2
                                    ↓
CO2R                               CO2R
 |                                  |
 CH————CH—CO2R  ←                  ·CH
   \   /                              \
    CH                                  CH—CO2R
    |                                 /
    CO2R                           ·CH
                                    |
                                   CO2R
```

For a number of unsaturated compounds, including particularly α,β-unsaturated carbonyl compounds, the mechanism above is supported by the fact that the pyrazolines may be isolated and their transformation to cyclopropane derivatives studied separately. It might be suggested that the transformations of pyrazolines to cyclopropanes involve an initial decomposition to an olefin and a diazo compound, a decomposition of the diazo compound to nitrogen and a methylene, and finally addition of the methylene to the olefin. In at least some instances, however, there is good evidence that this is not the case. Certain diaryldiazomethanes react with certain olefins to give cyclopropane derivatives under conditions where no decomposition of the diazo compound to a methylene would be expected. This behavior is explained by the pyrazoline mechanism (even though pyrazolines may not be isolable in these cases) on the basis of the great stability of the intermediate radical formed on loss of nitrogen.

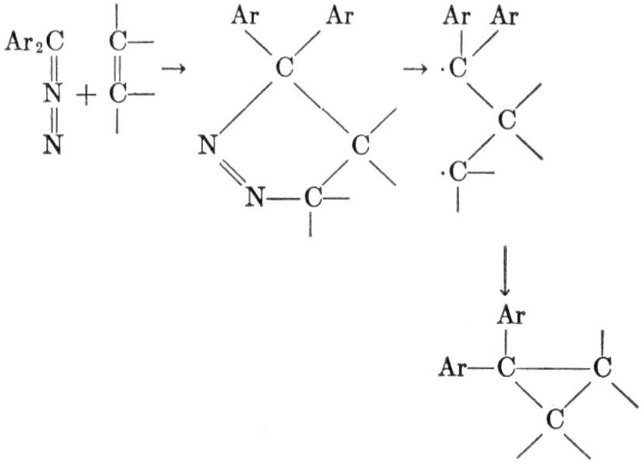

$$Ar_2C{=}N{=}N + {>}C{=}C{<} \rightarrow \text{(pyrazoline: } Ar_2C\text{, N=N, C, C ring)} \rightarrow Ar_2\dot{C}{-}C{<}{-}\dot{C}{<} \rightarrow Ar_2C{-}C{<}\text{(cyclopropane)}$$

Although it seems clear that some of the reactions in which three-membered rings are formed involve methylenes as intermediates and some involve addition by the diazo compound followed by loss of nitrogen, not enough mechanistic studies have been made to tell reliably what changes in reactant structures and reaction conditions are sufficient to change the mechanism from one type to another. In addition, it may be pointed out that in some cases neither of the two mechanisms just described will operate; for example, organocopper compounds may be the intermediates in certain copper-catalyzed reactions.

Reactions of Diazoacetic Esters with Olefins. Buchner and Geronimus found that the reaction of ethyl diazoacetate with styrene for about fourteen hours at 100° gives an ethyl phenylcyclopropanecarboxylate. Subsequent workers were able to isolate both geometrical isomers of the product and to show (e.g., by transformation of the phenyl group to a carboxy group) that the *cis* isomer is the principal product.

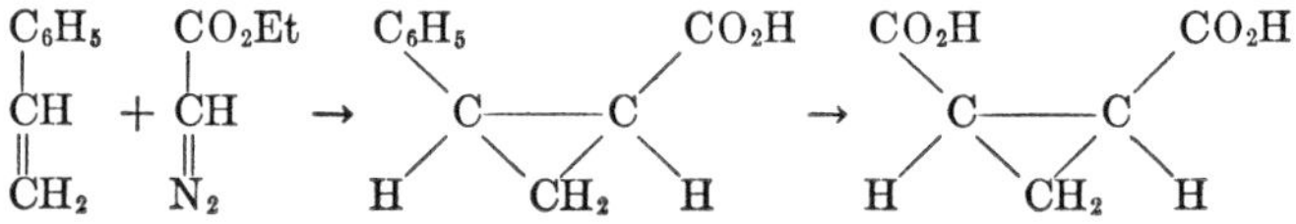

$$C_6H_5CH{=}CH_2 + N_2{=}CHCO_2Et \rightarrow (C_6H_5)(H)C{-}C(CO_2H)(H) \text{ [ring } CH_2\text{]} \rightarrow (CO_2H)(H)C{-}C(CO_2H)(H) \text{ [ring } CH_2\text{]}$$

Since it seems possible that the reaction conditions are sufficient to bring about the formation of carbethoxymethylene from the diazoacetic ester, it is plausible that the reaction involves the addition of a methylene to a double bond.

The addition of carbalkoxymethylenes to the 2-butenes has been shown to be stereospecifically *cis* (30, 172, 173). Doering and Mole found that the photolysis of methyl diazoacetate in the presence of *cis*-2-butene gives a mixture of two different methyl 1,2-dimethylcyclopropane-3-carboxylates whereas in the presence of *trans*-2-butene it gives a third isomeric methyl 1,2-dimethylcyclopropane-3-carboxylate.

$$H\text{—}\ddot{C}\text{—}CO_2CH_3 \xrightarrow{\textit{trans}\text{-2-butene}} \text{cyclopropane: } C(H)(CH_3)\text{—}C(CH_3)(H)\text{, ring } C(H)(CO_2CH_3)$$

$$\downarrow \textit{cis}\text{-2-butene}$$

$$\text{cyclopropane: } C(CH_3)(H)\text{—}C(CH_3)(H)\text{, ring } C(H)(CO_2CH_3) \; + \; \text{cyclopropane: } C(H)(CH_3)\text{—}C(H)(CH_3)\text{, ring } C(H)(CO_2CH_3)$$

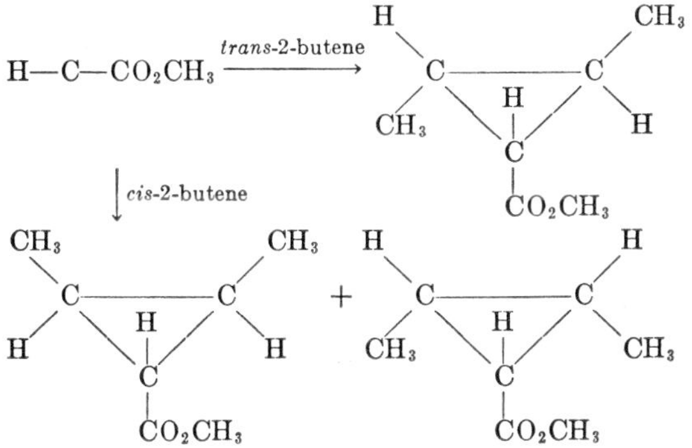

The reaction of *cis*-2-butene yields about two and one-half times as much of the *cis-trans* as of the all-*cis* product (172).

Dyakonov and coworkers resolved the acid obtained by hydrolysis of the ethyl 1,2-diphenylcyclopropane-3-carboxylate produced in the reaction of ethyl diazoacetate with *trans*-stilbene. This shows that the phenyl groups must be situated *trans* to each other on the cyclopropane ring (173) and thus confirms the stereospecifically *cis* nature of the addition of carbalkoxymethylenes to olefins.

There is strong evidence that the intermediates formed when diazoacetic esters are decomposed by the catalytic action of copper and its salts (and probably other metals and their salts also) are not the same as those formed by photolysis or pyrolysis. The intermediates formed by copper catalysis have little tendency to attack aromatic rings (especially those without strong electron-donor substituents) and carbon-hydrogen bonds. Instead, fumaric esters are usually produced, probably by the attack of the intermediate on a molecule of diazoester followed by loss of nitrogen. Doering and Mole found that in the reaction of butene with diazoacetic ester the yield of cyclopropane derivatives dropped from 39 per cent when ultraviolet light was used to 5–10 per cent when a copper-sulfate catalyst was used (172). Skell and Etter found that ethyl diazoacetate and cyclohexene give products VII, VIII, and IX in yields of 16, 10, and 21 per cent in the photolysis reaction, but that the yields are 69, 4, and 0 per cent, respectively, in the copper-catalyzed reaction (174).

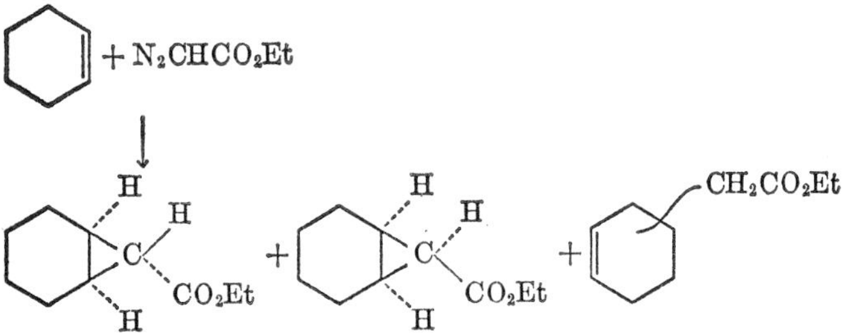

+ N_2CHCO_2Et

H, H, C, CO_2Et, H + H, H, C, CO_2Et, H + CH_2CO_2Et

VII VIII IX

Thus the copper intermediate appears to be much more discriminating, the photolysis intermediate giving almost random attack. Nevertheless, the copper intermediate showed little discrimination in its reaction with various pairs of olefins. The reactivity sequence: tetramethylethylene ~ trimethylethylene > cyclohexene > 1-hexene, speaks for the electrophilic character of the intermediate, but the first olefin in the sequence was only 1.8 times as reactive as the last one. Skell and Etter suggest that interactions between the alkyl groups of highly alkylated ethylenes and the bulky copper atom in the copper intermediate decrease the reactivity.

Addition of Diphenylmethylene to Olefins. Etter, Skovronek, and Skell found that the photolysis of diphenyldiazomethane in the presence of *cis*- and *trans*-2-butene gave mixtures of dimethyldiphenylcyclopropanes in each case, the product ratios being 15:1 and 2.3:1 when *cis*- and *trans*-2-butene, respectively, were the reactants* (175). They proposed that the diphenylmethylene is a triplet, which adds to the butenes to give a diradical that does not cyclize fast enough to maintain its stereochemical integrity.

CH₃ CH₃ CH₃ CH₃

$C_6H_5—\ddot{C}—C_6H_5$ + (CH₃)(H)C=C(CH₃)(H) → (CH₃)(H)Ċ—C(CH₃)(H)—ĊH(C₆H₅)₂

H H H ·C H

C₆H₅ C₆H₅

H CH₃ H CH₃ CH₃ CH₃

C——C C—C C——C

CH₃ C H ← CH₃ ·C H H C H

C₆H₅ C₆H₅ C₆H₅ C₆H₅ C₆H₅ C₆H₅

Rotation around the carbon-carbon single bonds of the diradical is not fast compared to ring closure. If it were, the same product mixture should be obtained from *cis*- and *trans*-2-butene. The evidence for non-stereospecificity in the addition of diphenylmethylene to the 2-butenes is weakened by the fact that the reaction products were not shown to be stereochemically stable under the reaction conditions. The neces-

* Closs and Closs agreed that the reaction is not completely stereospecific but they obtained rather different quantitative results. The reactions of *cis*- and *trans*-2-butene were found to be 87 per cent and 96 per cent stereospecific, respectively (176).

sity of such a check experiment is shown by the observation of Doering and Jones that the 2,3-dimethylspiro(cyclopropane-1,9[1]-fluorene)'s are subject to light-catalyzed *cis-trans* isomerization (175a).

However, the triplet (diradical) character of diphenylmethylene is also supported by the fact that 1,3-butadiene and 1,1-diphenylethylene are more than 100 times as reactive as isobutylene, 1-hexene, or cyclohexene toward diphenylmethylene. This is a typical radical reactivity sequence.

Having obtained some of the first chemical evidence for a divalent carbon derivative existing in the triplet form, Skell and coworkers suggested that the term "carbenes" be used for singlets and "methylenes" be used for triplets. This suggestion has not yet been used widely.

Closs and Closs were able to combine the evidence described above with additional observations of their own to show that the elements of diphenylmethylene may be added to a double bond without diphenylmethylene itself being necessarily a real intermediate in the reaction. They observed that the reaction of methyllithium with diphenyldibromomethane in the presence of olefins (including the 2-butenes) yields 1,1-diphenylcyclopropane derivatives with *complete* stereospecificity (176). They therefore concluded that it is diphenylbromomethyllithium that adds to olefins under their reaction conditions.

$$(C_6H_5)_2CBr_2 \xrightarrow{CH_3Li} (C_6H_5)_2C(Li)\text{—}Br$$

$$>C{=}C< \; + \; (C_6H_5)_2C(Li)Br \rightarrow \text{(cyclopropane: } >C\text{—}C<\text{ bridged by } C(C_6H_5)_2) \; + \; LiBr$$

There is some possibility that it is singlet diphenylmethylene that was responsible for this stereospecific addition reaction. Nevertheless it seems clearly established that diphenylmethylene is a triplet in the ground state; electron-paramagnetic-resonance measurements by two different groups of workers support this point (163a, 176a, 176b). These EPR measurements are said to rule out a diphenylmethylene structure with the two aromatic rings perpendicular and the two bonds to the divalent carbon collinear, the structure that had been thought to be most plausible for triplet diphenylmethylene.

Internal Additions of Methylenes to Double Bonds. Closs and Closs showed that alkenylmethylenes are transformed spontaneously to cyclopropene derivatives, a reaction that can be regarded as an internal addition of a methylene to a carbon-carbon double bond. The intermediate 2,3-dimethylpropenylmethylene was generated not only from the diazo compound but also by the α-dehydrochlorination of 1-chloro-2,3-dimethyl-2-butene and by the reaction of 2,3-dimethylpropenyllithium with methylene chloride (177).

$$CH_2Cl_2 \xrightarrow{RLi} H\text{—}\ddot{C}\text{—}Cl$$

$$(CH_3)_2C\text{=}C(CH_3)\text{—}CH\text{=}NNHSO_2C_7H_7 \xrightarrow{base} (CH_3)_2C\text{=}C(CH_3)CHN_2$$

$$H\text{—}\ddot{C}\text{—}Cl \xrightarrow{(CH_3)_2C=C(Li)CH_3} (CH_3)_2C\text{=}C(CH_3)\ddot{C}\text{—}H$$

$$(CH_3)_2C\text{=}C(CH_3)CHN_2 \longrightarrow (CH_3)_2C\text{=}C(CH_3)\ddot{C}\text{—}H$$

$$(CH_3)_2C\text{=}C(CH_3)CH_2Cl \xrightarrow{n\text{-BuLi}} (CH_3)_2C\text{=}C(CH_3)\ddot{C}\text{—}H$$

$$(CH_3)_2C\text{=}C(CH_3)\ddot{C}\text{—}H \longrightarrow (CH_3)_2\overset{}{C}\text{—}C(CH_3)\text{=}CH \text{ (cyclopropene ring)}$$

Larger rings may also be formed by internal addition to double bonds, as Stork and Ficini found in the decomposition of the diazo ketone shown below to give bicyclo[4.1.0]-2-heptanone (178).

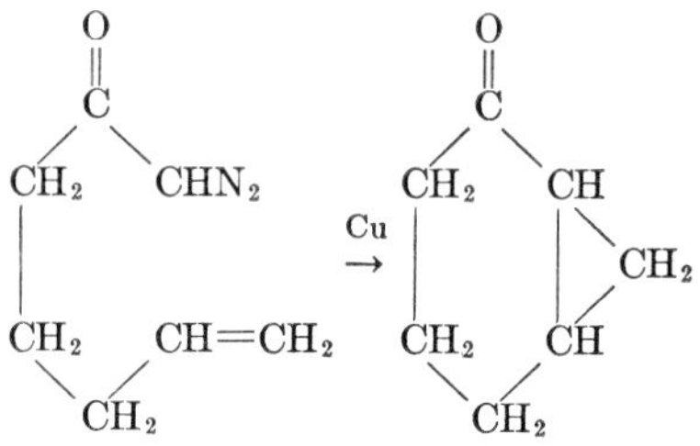

$$\begin{array}{c}\text{O} \\ \| \\ \text{C} \\ \diagup \quad \diagdown \\ \text{CH}_2 \qquad \text{CHN}_2 \\ | \\ \text{CH}_2 \qquad \text{CH}=\text{CH}_2 \\ \diagdown \quad \diagup \\ \text{CH}_2\end{array} \xrightarrow{\text{Cu}} \begin{array}{c}\text{O} \\ \| \\ \text{C} \\ \diagup \quad \diagdown \\ \text{CH}_2 \qquad \text{CH} \\ | \qquad\qquad | \diagdown \\ | \qquad\qquad | \;\; \text{CH}_2 \\ | \qquad\qquad | \diagup \\ \text{CH}_2 \qquad \text{CH} \\ \diagdown \quad \diagup \\ \text{CH}_2\end{array}$$

1,3-Dipolar Addition of Acylmethylenes to Multiple Bonds. Huisgen and coworkers pointed out that acylmethylenes are among the species that should take part (as 1,3-dipoles) in 1,3-dipolar addition reactions (179). Such an addition can be written in general terms as follows (assuming that the acylmethylene is a singlet):

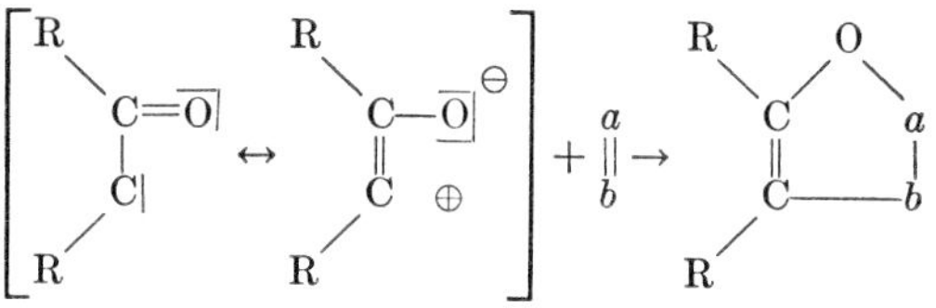

$$\left[\begin{array}{c}\text{R} \\ \diagdown \\ \text{C}=\overline{\text{O}}| \\ | \\ \text{C}| \\ \diagup \\ \text{R}\end{array} \leftrightarrow \begin{array}{c}\text{R} \\ \diagdown \\ \text{C}-\overline{\text{O}}|^{\ominus} \\ \| \\ \text{C}\;\oplus \\ \diagup \\ \text{R}\end{array}\right] + \begin{array}{c}a \\ \| \\ b\end{array} \rightarrow \begin{array}{c}\text{R} \qquad \text{O} \\ \diagdown \;\; \diagup \quad \diagdown \\ \text{C} \qquad\quad a \\ \| \qquad\qquad | \\ \text{C}\text{———}b \\ \diagup \\ \text{R}\end{array}$$

Attempts to find products of 1,3-dipolar addition by benzoylphenylmethylene generated by the decomposition of $C_6H_5COC(N_2)C_6H_5$ were unsuccessful, perhaps because the intermediate methylene rearranged to diphenylketene faster than it added to a double bond or perhaps because no methylene was formed (the rearrangement may have been concerted, with the phenyl group migrating at the same time that nitrogen was lost).

In order to study an acylmethylene with less tendency to rearrange, the decomposition of tetrachloro-α-diazocyclohexadienone (X) was studied in the presence of several unsaturated compounds. The methylene intermediate in this case (XI) should possess some aromatic resonance stabilization, which should decrease its tendency to rearrange.

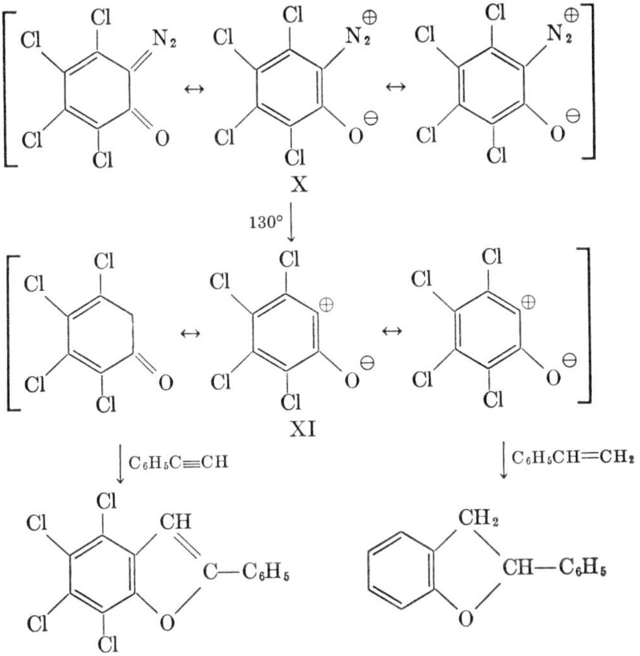

X

130°

XI

$C_6H_5C{\equiv}CH$ $\qquad$ $C_6H_5CH{=}CH_2$

As shown above, the decomposition of this diazo ketone at 130° in the presence of styrene or phenylacetylene leads to the formation of a product containing a new five-membered ring. When the diazo ketone X is decomposed in benzonitrile solution, either thermally at 130° or photochemically at 20°, 4,5,6,7-tetrachloro-2-phenylbenzoxazole is formed. In several cases the rate of disappearance of X was found to be independent, to a first approximation, of the nature of the solvent or the unsaturated compound present. This shows that the reactions all proceed by the same rate-controlling step and therefore not by attack of the unsaturated compound on the diazo ketone. An exception to this generalization was found when the unsaturated compound is diphenylketene, which brings about the decomposition of X even at room temperature. The reaction in this case does not involve a methylene but is initiated by direct attack of the ketene on X.

A study of the generation of the methylene XI in the presence of the dimethyl esters of maleic and fumaric acids showed that the ester of a *trans*-dicarboxylic acid was obtained in both cases.

Cl Cl Cl Cl N_2 O + CO_2CH_3 CH CH CO_2CH_3 → Cl Cl Cl Cl H CO_2CH_3 C CO_2CH_3 C H O

The possibility that the maleic ester isomerized before it reacted or that the *cis*-dicarboxylic ester product isomerized after it was formed has not yet been eliminated, however.

Probably one of the most important reasons why the acylmethylene XI gives 1,3-dipolar addition rather than 1,1-addition like many methylenes, is the fact that 1,3-addition preserves the aromatic ring present in XI.

Huisgen and coworkers pointed out that another acylmethylene that should show relatively little tendency to rearrange is carbethoxymethylene, which on rearrangement would lose the internal resonance stabilization of the carbethoxy group. Unfortunately, this internal resonance stabilization is also lost upon 1,3-dipolar addition and, perhaps for this reason, carbethoxymethylene usually gives three-membered rings rather than five-membered rings on addition to carbon-carbon multiple bonds. With carbon-nitrogen triple bonds, however, 1,3-dipolar addition, to give a resonance-stabilized heteroaromatic compound, has been observed.

CHN_2 C EtO O $\xrightarrow{145°}$ H C C EtO O $\xrightarrow{C_6H_5CN}$ N C_6H_5 CH C C O EtO

Another way in which the products to be expected from 1,3-dipolar addition have been obtained in several cases is the decomposition of the diazo ketone or diazo ester in the presence of a copper catalyst. In these cases, of course, there is considerable doubt that methylenes are really reaction intermediates.

Miscellaneous Additions of Methylenes to Carbon-Carbon Multiple Bonds. A number of derivatives of cyclopropane have been prepared by the decomposition of diazo compounds in the presence of olefins. In a number of cases, such as the decomposition of diazoacetone in the presence of various olefins (180), copper catalysts were used and therefore the reactions may well have involved organocopper intermediates rather than methylenes. In other cases, such as the photolysis of ethyl trifluoroacetyldiazoacetate in the presence of cyclohexene (171), a methylene intermediate seems more plausible.

Jones found that the treatment of 2,2-diphenylcyclopropylnitrosourea (XII) with base in the presence of olefins leads to the stereospecific formation of spiropentane derivatives, e.g., (181)

$$\underset{\text{XII}}{(C_6H_5)_2\overset{\diagdown}{C}-\!\!\underset{CH_2}{\;}\!\!-\overset{\diagup}{C}HN(NO)-CONH_2} \xrightarrow{\text{base}} (C_6H_5)_2C\text{—}CN_2 \text{ (ring via } CH_2)$$

$$\downarrow$$

$$\text{spiro: } (C_6H_5)_2C(CH_2)C\text{[}C(H)\text{—}CH_3\text{][}C(H)\text{—}CH_3\text{]} \xleftarrow{cis\text{-2-butene}} (C_6H_5)_2C\text{—}C \text{ (ring via } CH_2)$$

The reaction also gives 1,1-diphenylallene (cf. p. 160), whose yield in the absence of olefin approaches the quantitative.

The diazo compound is too stable to be isolated even at −5 to 0° but quantitative studies on the effect of reactant concentrations on product yields show that the allene has two precursors, only one of which can add to olefins to give spiropentane derivatives. Apparently some of the diazo compound decomposes directly to diphenylallene and some of it yields 2,2-diphenylcyclopropylidene, which may either add to olefin or rearrange to the allene (181a).

Propargylene ($H{-}C{\equiv}C{-}\dot{\underset{\cdot}{C}}{-}H$) is a methylene that, in view of its reported non-stereospecific addition to olefins, appears to be a triplet. Skell and Klebe found that the photolysis of diazopropyne in the presence of *cis*- and *trans*-2-butene gave, in each case, a mixture (but not the same mixture) of the three isomeric 1,2-dimethyl-3-ethynylcyclopropanes (182).

$$CH{\equiv}C{-}CHN_2 \xrightarrow{h\nu} [H{-}C{\equiv}C{-}\dot{\underset{\cdot}{C}}{-}H \leftrightarrow H{-}\dot{C}{=}C{=}\dot{C}{-}H \leftrightarrow \text{etc.}]$$

$$\downarrow \text{2-butene}$$

(H, CH_3)C—C(H, CH_3) cyclopropane with ring C(H)—C≡CH + (H, CH_3)C—C(CH_3, H) cyclopropane with ring C(H)—C≡CH + (CH_3, H)C—C(H, CH_3) cyclopropane with ring C(H)—C≡CH

It seems particularly probable that propargylene should be a triplet in view of the stability that should be associated with a linear C_3H_2 molecule having three *sp*-hybridized carbon atoms surrounded by a cylindrically symmetrical sheath of six electrons.

Addition of Methylenes to Carbon-Carbon Triple Bonds. Dyakonov and Komendantov found that ethyl diazoacetate reacts with 1-phenylpropyne in the presence of anhydrous copper sulfate to give the ethyl ester of a methylphenylcyclopropenecarboxylic acid (183). However, because of the nature and specificity of the catalyst (with copper-bronze only ethyl fumarate was obtained), the intermediacy of a free methylene is in considerable doubt. Methylene intermediates seem more plausible in Breslow's thermal reaction of phenyldiazoacetonitrile with diphenylacetylene to give triphenylcyanocyclopropene, a precursor of the triphenylcyclopropenyl cation, the first aromatic species known with only two pi electrons (184).

$$C_6H_5C{\equiv}CC_6H_5 + C_6H_5\underset{\underset{N_2}{\|}}{C}CN \rightarrow \underset{C_6H_5C{=\!=}CC_6H_5}{\overset{C_6H_5{-}C{-}CN}{\diagup\;\diagdown}}$$

$$\downarrow BF_3, H_2O$$

$$\left[\text{etc.} \leftrightarrow \quad \underset{C_6H_5C{-\!-}\overset{\oplus}{C}C_6H_5}{\overset{\overset{C_6H_5}{|}}{\overset{C}{/\!/\;\diagdown}}} \quad \leftrightarrow \quad \underset{C_6H_5C{=\!=}CC_6H_5}{\overset{\overset{C_6H_5}{|}}{\overset{C\oplus}{\diagup\;\diagdown}}}\right]$$

Breslow and coworkers also prepared cyclopropene derivatives by the decomposition of diazomalonic ester and diazoacetic ester in the presence of acetylenes, and Doering and Mole found that the photolysis of methyl diazoacetate in the presence of 2-butyne yields methyl 1,2-dimethylcyclopropene-3-carboxylate (172).

REACTIONS OF METHYLENES WITH COMPOUNDS WITH UNSHARED ELECTRON PAIRS

Since various methylenes have been found in a number of cases (e.g., addition to olefins) to be rather strongly electrophilic species it is not surprising that methylenes often appear to coordinate with the unshared electron pairs of various Lewis bases. This type of reaction occurs most readily when the negative charge that is thus placed on the formerly divalent carbon atom is stabilized by electron-withdrawing substituents.

Reactions of Methylenes with Organic Oxygen Compounds. Although a methylene intermediate may be written for the decomposition reactions of various diazo compounds in alcohol solution to yield the corresponding ethers,

$$\begin{array}{ccc} R_2CN_2 & \xrightarrow{-N_2} & R_2C \\ \downarrow{\scriptstyle R'OH} & & \downarrow{\scriptstyle R'OH} \\ R_2CHN_2^+ & \rightarrow & R_2CHOR' \end{array}$$

there appear to be no careful studies of the reaction in which an alternative intermediate diazonium cation can be ruled out. With ethers the intermediacy of a diazonium cation is less plausible. The observation by Gutsche and Hillman that the decomposition of ethyl diazoacetate in 2-phenyl-1,3-dioxolane solution gives ethyl 3-phenyl-2-p-dioxanecarboxylate seems reasonably explained by their proposed reaction mechanism (185).

$$\text{(2-phenyl-1,3-dioxolane: } CH_2\text{–}CH_2 \text{ with O–}CHC_6H_5\text{–O)} \xrightarrow{H\text{—}C\text{—}CO_2Et} \text{(}CH_2\text{–}CH_2\text{, }\overset{\oplus}{O}\text{—}\overset{\ominus}{C}HCO_2Et\text{, }CHC_6H_5\text{, O)} \rightarrow \text{(}CH_2\text{–}CH_2\text{, O, }CHCO_2Et\text{–}CHC_6H_5\text{, O)}$$

It also seems possible, though, that the formation of the zwitterionic intermediate shown above is followed by an S_N1-type ring opening and reclosure. The stability of the intermediate carbonium ion thus formed would offer one explanation (there are others possible) for the somewhat different behavior of simple dialkyl and arylalkyl ethers. Di-n-butyl ether, for example, was found to react with ethyl diazoacetate at 140° to give 1-butene and ethyl n-butoxyacetate in addition to the products of C—H insertion reactions (186).

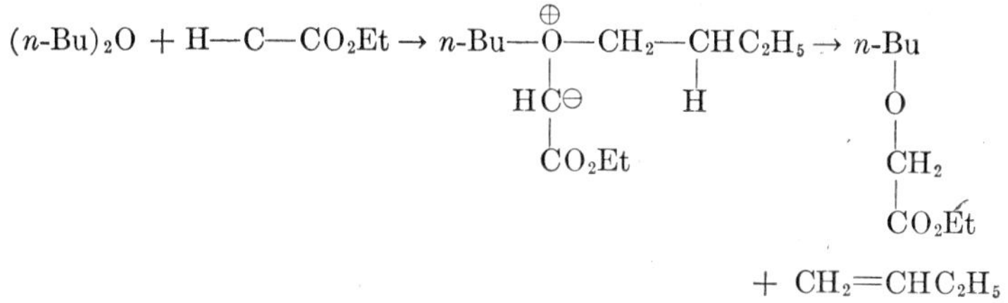

$$(n\text{-Bu})_2O + H\text{—}C\text{—}CO_2Et \rightarrow n\text{-Bu}\text{—}\overset{\oplus}{O}(\text{—}CH_2\text{—}CH(H)C_2H_5)(\text{—}HC^{\ominus}\text{—}CO_2Et) \rightarrow n\text{-Bu—O—}CH_2\text{—}CO_2Et + CH_2{=}CHC_2H_5$$

Several arylalkyl ethers were found to react under similar conditions to give aryloxyacetate esters, but the fate of the alkyl part of the original ether was not determined.

Cyclohexanone reacts with ethyl diazoacetate in the presence or absence of copper powder to give an unsaturated ether-ester, among other products (187).

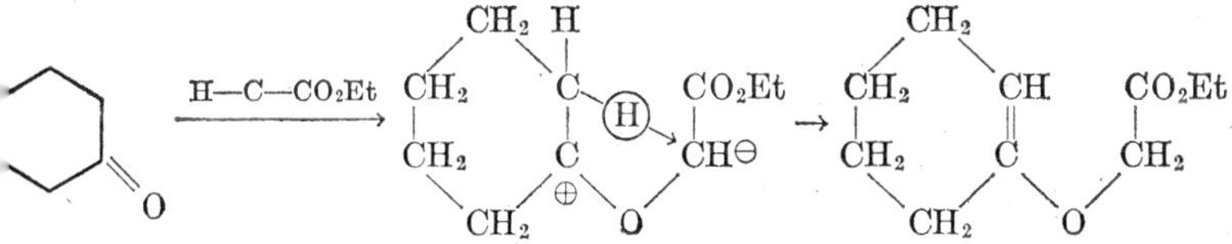

$$\text{cyclohexanone} \xrightarrow{H\text{—}C\text{—}CO_2Et} \text{[ring: } CH_2, CH_2, CH_2, CH_2, CH(H), \overset{\oplus}{C}\text{—}O\text{—}CH^{\ominus}\text{—}CO_2Et] \rightarrow \text{[ring: } CH_2, CH_2, CH_2, CH_2, CH{=}C\text{—}O\text{—}CH_2\text{—}CO_2Et]$$

Kharasch and coworkers suggested a somewhat different mechanism for the reaction, but, as with many of the reactions described in this chapter, there is little evidence that can be used to distinguish between the two possibilities.

The copper-catalyzed decomposition of various diazo compounds in the presence of furan and benzofuran derivatives has been found to give ring-opening and addition reactions; phenol undergoes attack on oxygen and on the ortho position of the ring.

Reactions of Methylenes with Amines. Aniline reacts on heating with various diazo compounds to yield *N*-substituted anilines. With benzyldimethylamine at 150°, 9-diazofluorene gives 9-benzyl-9-dimethylaminofluorene, presumably via an intermediate zwitterion that undergoes a Stevens-type rearrangement.

$$\text{fluorenylidene-}N_2 \rightarrow \text{fluorenylidene (C:)} \xrightarrow{C_6H_5CH_2NMe_2} C_6H_5CH_2\overset{\oplus}{N}Me_2\text{—}C^{\ominus}(\text{fluorenyl}) \rightarrow C_6H_5CH_2\text{—}C(NMe_2)(\text{fluorene})$$

Kirmse, Horner, and Hoffman found that the photolysis of diphenyldiazomethane in the presence of diethylamine, isopropylamine, or acetamide results in the insertion of the benzhydryl group into the N—H bonds (163).

Franzen and Kuntze studied the photolysis of ethyl diazoacetate in the presence of several amines (188). Triethylamine gave the products to be expected from insertion at the α- and β-carbon-hydrogen bonds and also some ethyl α-diethylaminoacetate. It seems plausible that the latter product arose from attack of carbethoxymethylene on the nitrogen atom of the amine, as shown below.

$$(C_2H_5)_3N + H\text{—}C\text{—}CO_2Et \rightarrow (C_2H_5)_2\overset{\oplus}{N}(\overset{\ominus}{C}H\text{—}CO_2Et)\text{—}CH_2\text{—}CH_2\text{—}H$$

$$\downarrow$$

$$(C_2H_5)_2N\text{—}CH_2\text{—}CO_2Et + CH_2{=}CH_2$$

Reactions of Methylenes with Organic Halides. Dyakonov and Vinogradova found that the copper-catalyzed decomposition of ethyl diazoacetate in the presence of allyl halides gives unsaturated haloesters as shown below.

$$CH_2{=}CHCH_2Cl + N_2CHCO_2Et \rightarrow CH_2{=}CHCH_2\underset{\displaystyle Cl}{\underset{|}{C}}HCO_2Et$$

This observation and Phillips' report that ethyl 2,5-dichloro-4-pentenoate is formed from 3,3-dichloropropene and ethyl diazoacetate (189) may be rationalized in terms of a mechanism involving the intermediate formation of a chloronium ion.

$$CH_2{=}CHCHCl_2 \xrightarrow{H-C-CO_2Et} \begin{array}{l} Cl \\ | \\ CH-\overset{\oplus}{Cl}-\overset{\ominus}{C}HCO_2Et \rightarrow \\ | \\ CH \\ \| \\ CH_2 \end{array}$$

$$ClCH{=}CHCH_2\underset{\displaystyle Cl}{\underset{|}{C}}HCO_2Et$$

However, this mechanism assigns no role to the copper catalyst.

A similar mechanism could be used to explain the observations by Urry and Wilt that the photolysis of methyl diazoacetate in the presence of chloroform and bromotrichloromethane leads to the formation of methyl α,β,β-trichloropropionate and methyl β-bromo-α,β,β-trichloropropionate, respectively (among other products), reactions in which the intermediacy of a methylene seems more probable (190). However, there is also the possibility that the reactions are one-step insertions of carbomethoxymethylene into a carbon-chlorine bond. Still another mechanism that is in agreement with the experimental observations was suggested by Urry and Wilt. For the reaction with chloroform this mechanism is the following:

$$CHCl_3 + H—C—CO_2Me \rightarrow \cdot CCl_3 + \cdot CH_2CO_2Me$$

$$\cdot CCl_3 + N_2CHCO_2Me \rightarrow Cl_3C—\dot{C}H—CO_2Me$$

$$Cl_3C—\dot{C}H—CO_2Me \rightarrow Cl_2\dot{C}—CHCl—CO_2Me$$

$$Cl_2\dot{C}—CHCl—CO_2Me + CHCl_3 \rightarrow Cl_2CH—CHCl—CO_2Me + \cdot CCl_3$$

REARRANGEMENTS OF METHYLENES

The Wolff Rearrangement (191). Wolff, and subsequently many other workers, showed that α-diazoketones may rearrange in the presence of water to give carboxylic acids, in the presence of ammonia to give amides, and in the presence of alcohols to give esters. All of these products would be expected to result from the further reaction of an intermediate ketene and indeed in inert solvents ketenes have been isolated.

$$R—\overset{\overset{O}{\|}}{C}—CH{=}N_2 \rightarrow R—\overset{\overset{O}{\|}}{C}—\ddot{C}—H \qquad RCH_2CO_2R'$$

$$\downarrow \qquad \nearrow^{R'OH}$$

$$RCH_2CO_2H \xleftarrow{H_2O} R—CH{=}C{=}O \xrightarrow{NH_3} RCH_2CONH_2$$

The mechanism above seems plausible for the thermal or photochemical (192) Wolff rearrangement, but for the reaction as ordinarily carried out, in the presence of a silver catalyst, it is incomplete since it assigns no role to the silver. It has been suggested that the silver is necessary for the rearrangement of the intermediate methylene since in the absence of silver large amounts of ketol ($RCOCH_2OH$) are obtained (in aqueous solution), but it seems more plausible that the silver acts by speeding the decomposition of the diazo compound so that it can compete more favorably with a simultaneous polar reaction with water leading to the ketol.

The Wolff rearrangement did not become a reaction of major importance until Arndt and Eistert showed that diazomethyl ketones could be prepared quite generally by the reaction of acid halides with diazomethane (191).

$$RCOCl + 2CH_2N_2 \rightarrow RCOCHN_2 + CH_3Cl + N_2$$

The reactions of acid halides with diazo compounds other than diazomethane have received relatively little attention.

Silver-catalyzed Wolff rearrangements will not be discussed in detail here because they may not involve methylenes and they have been reviewed several times elsewhere (191).

Schroeter found that simply on heating to about 60° azibenzil (phenyl α-diazobenzyl ketone) loses nitrogen and rearranges to diphenylketene. Horner and coworkers found that ultraviolet light brings about the same reaction at 0° (192). If benzoylphenylmethylene is really an intermediate in this reaction, that is, if the migration of a phenyl group follows, rather than accompanies, the loss of nitrogen, it would be interesting to learn whether the methylene may form diphenylacetylene oxide occasionally before it rearranges. Franzen pointed out that such a reaction would make the two acyclic carbon atoms equivalent. He disposed of the possibility by showing that azibenzil labelled in the carbonyl group with ^{14}C yields diphenylketene labelled only in the carbonyl group (193).

$$C_6H_5-\overset{N_2}{\overset{\|}{C}}-\overset{O}{\overset{\|}{\overset{*}{C}}}-C_6H_5 \rightarrow C_6H_5-\ddot{C}-\overset{O}{\overset{\|}{\overset{*}{C}}}-C_6H_5 \rightarrow (C_6H_5)_2C{=}\overset{*}{C}{=}O$$

$$\not\updownarrow$$

$$C_6H_5-C\overset{O}{=\!=}\overset{*}{C}-C_6H_5$$

The photolysis of α-diazoketones has been used to bring about ring contraction to yield compounds with small strained rings and also medium-sized rings. Of particular interest is the photolysis of diazocamphor, which was found by Horner and Spietschka to yield a derivative of bicyclo[2.1.1]hexane (194).

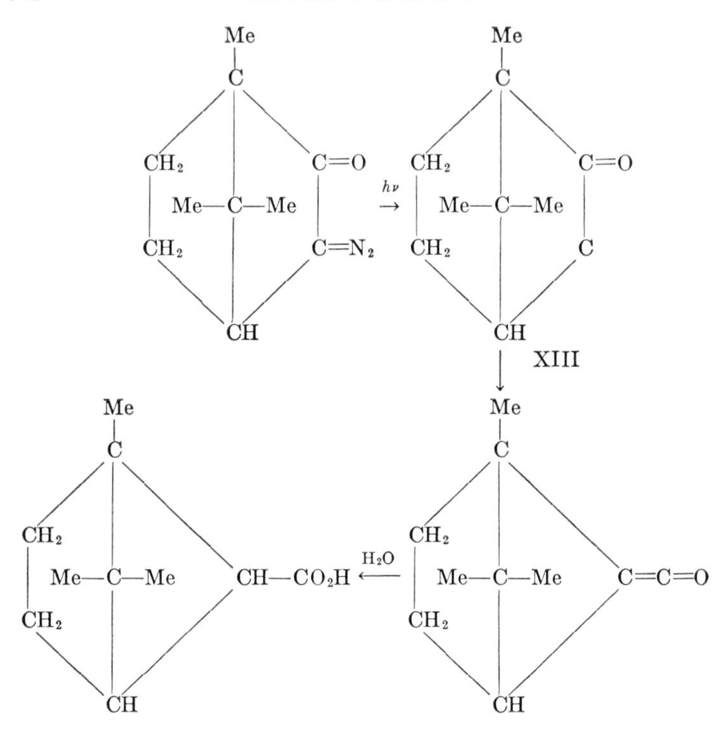

The methylene XIII would seem to be a plausible intermediate in this reaction, except that this intermediate has already been written on p. 110 as the precursor of dehydrocamphor, an internal-insertion product resulting from the thermal decomposition of diazo camphor at 140°. It is possible that the methylene undergoes both of the reactions suggested, with internal insertion being the predominant reaction at higher temperatures and rearrangement predominant at room temperature. There are other possible complications that might bear investigation, however. For example, although the dehydrocamphor whose structure was proved by Bredt and Holz appears to be the same as the material prepared earlier by Schiff by the *thermal* decomposition of diazocamphor, the material used by Bredt and Holz was prepared by the copper-catalyzed reaction. Horner and Spietschka suggested that in the formation of dehydrocamphor the diazocamphor first rearranges to a tricyclic azo compound. It also seems possible that the difference between the two reactions may be attributable to a difference in the amount of excess energy possessed by the methylene formed in the two different ways.

The diazo anhydrides (*o*-quinone diazides) resulting from the diazotization of *o*-aminophenols may be seen, from an examination of their various contributing resonance structures, to be α-diazo ketones. Therefore the photochemical transformation of such diazo anhydrides to derivatives of cyclopentadienecarboxylic acid and products of its further reaction, discovered by Süs, may be considered a type of Wolff rearrangement (195).

$$\left[O^{\ominus}\text{-}C_6H_4\text{-}N_2^{\oplus} \leftrightarrow O{=}C_6H_4{=}N_2 \right] \xrightarrow{h\nu} \text{O=C(CH=CH-CH=C}\text{-)} $$

$$\downarrow$$

$$\text{CH=CH-CH=CH-CH(CO}_2\text{H)} \xleftarrow{H_2O} \text{CH=CH-CH=CH-C=C=O}$$

The reaction may be applied to polycyclic diazo anhydrides, naphthalene derivatives yielding indene derivatives, and fluorene derivatives being formed from phenanthrene derivatives. Heterocyclic compounds may also be used, as shown below in the synthesis of an indole carboxylic acid from the corresponding quinoline diazo anhydride.

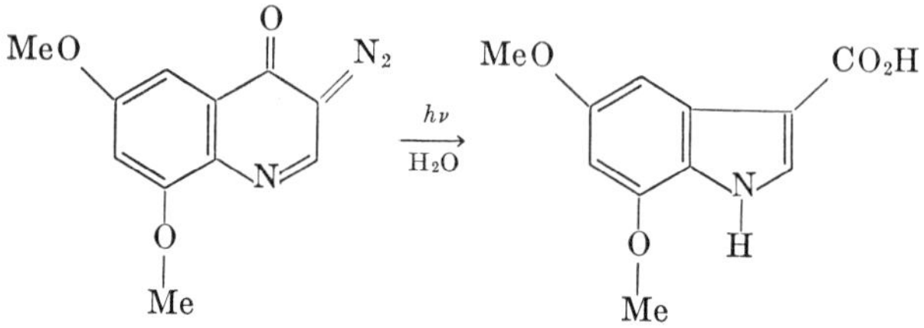

MeO O N$_2$ N O Me $\xrightarrow[H_2O]{h\nu}$ MeO CO$_2$H N H O Me

The reaction may be brought about by heating the diazo anhydrides as well as by photolyzing them.

In several cases ketene acetals or unsaturated derivatives of γ-butyrolactone are formed, probably by reaction of the diazo anhydride with the ketene formed by rearrangement of the intermediate methylene.

O N$_2$ $h\nu$ → O—C O

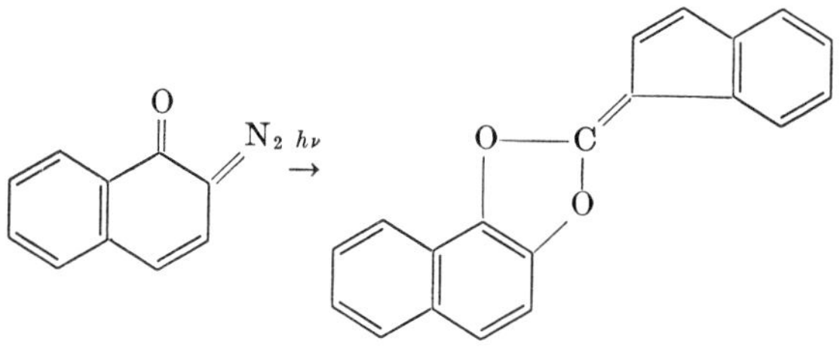

It would be interesting to learn whether the two substituted carbon atoms in diazo anhydrides become equivalent during the photolysis reaction. It seems more likely that they should than that the two acyclic carbon atoms in azibenzil (193) should, since a diazo anhydride could form a resonance-stabilized benzyne oxide.

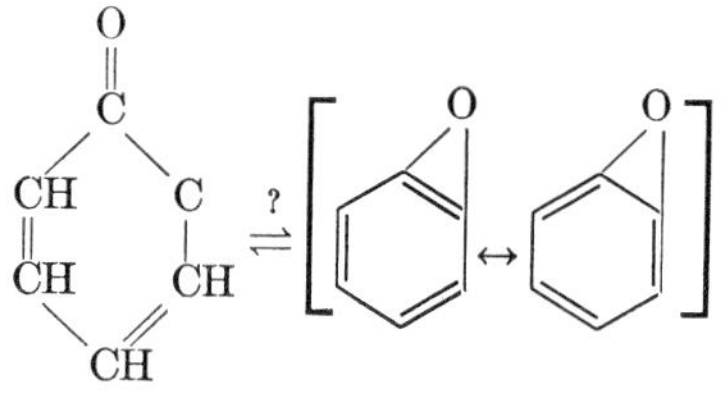

O C CH C CH CH CH ? ⇌ [O ↔ O]

Other Rearrangements of Acylmethylenes. In several cases the decomposition of α-diazoketones, presumably to give intermediate methylenes, has been found to give rearrangement products other than those to be expected from the Wolff rearrangement. Newman and Arkell, for example, found that the rearrangement of 4-diazo-2,2,5,5-tetramethyl-3-hexanone gives only 0–3 per cent of the expected Wolff-rearrangement product di-*t*-butylketene, presumably because of the difficulty of joining two bulky *t*-butyl groups to the same carbon atom (196). The principal product of the decomposition (either thermal or photochemical) of the diazoketone is 2,2,4,5-tetramethyl-4-hexen-3-one, the result of the migration of a methyl group in the intermediate methylene.

$$(CH_3)_3C{-}C({=}O){-}C({=}N_2){-}C(CH_3)_3 \rightarrow (CH_3)_3C{-}C({=}O){-}\ddot{C}{-}C(CH_3)_3 \rightarrow (CH_3)_3C{-}C({=}O){-}C(CH_3){=}C(CH_3)_2$$

Franzen studied the decomposition of seven α-diazo ketones of the type $Ar{-}CO{-}CN_2{-}CH_2R$ (where R may be H) in the presence of silver oxide and found the α,β-unsaturated ketones ($ArCOCH{=}CHR$) to be the reaction products in yields of 69–92 per cent. Since this reaction may well have involved an organosilver compound as an intermediate, it may be more relevant to the chemistry of methylenes to point out that three additional α-diazo ketones were found to give α,β-unsaturated ketones in yields of 50 to 68 per cent when decomposed by a silver-oxide catalyst *or by ultraviolet light* (197). In the photolytic reaction (but not in the silver oxide-catalyzed reaction) the carboxylic acid to be expected from the Wolff rearrangement was found in two cases.

$$C_6H_5CH_2{-}CN_2{-}COCH_3 \xrightarrow[H_2O]{h\nu} \begin{array}{l} 50\%\ C_6H_5CH{=}CHCOCH_3 \\ 17\%\ C_6H_5CH_2CH(CH_3)CO_2H \end{array}$$

$$(CH_3)_2CHCH_2{-}CN_2{-}COCH_3 \xrightarrow[H_2O]{h\nu} 54\%\ (CH_3)_2CHCH{=}CHCOCH_3$$

$$n\text{-}C_3H_7CH_2{-}CN_2{-}COCH_3 \xrightarrow[H_2O]{h\nu} \begin{array}{l} 60\%\ n\text{-}C_3H_7CH{=}CHCOCH_3 \\ 23\%\ n\text{-}C_3H_7CH_2CH(CH_3)CO_2H \end{array}$$

Franzen pointed out that in reactions carried out at higher temperatures (his were carried out at approximately room temperature) the yield of acid (or ester in the presence of alcohol) increases and that of α,β-unsaturated ketone appears to decrease. He suggested that near room temperature those intermediate methylenes that can undergo 1,2-hydrogen shifts to give α,β-unsaturated ketones will do so predominantly. Diazomethyl ketones cannot do so and therefore undergo the Wolff rearrangement. The photolysis of 3-diazonopinone (XIV) and a number of related ring-contraction reactions are exceptions to this generalization, however (198).

$$\underset{\text{XIV}}{(CH_3)_2C\begin{cases} CH-C{=}O \\ CH_2 \quad C{=}N_2 \\ CH-CH_2 \end{cases}} \xrightarrow[H_2O]{h\nu} (CH_3)_2C\begin{cases} CH-CHCO_2H \\ CH_2 \\ CH-CH_2 \end{cases}$$

The factors that determine whether the intermediate acylmethylene rearranges to a ketene or an α,β-unsaturated ketone are apparently not completely understood.

As Franzen pointed out, the reactions of dialkylacetylenes with peracids yield α,β-unsaturated ketones and the acids to be expected from ketene hydration, and therefore acylmethylenes seem plausible intermediates (197).

$$R-C{\equiv}C-CH_2R' \xrightarrow{R\overset{O}{\overset{\|}{C}}OOH} R-\overset{O}{\overset{\|}{C}}-\ddot{C}-CH_2R'$$

$$R-\overset{O}{\overset{\|}{C}}-\ddot{C}-CH_2R' \longrightarrow R-\overset{O}{\overset{\|}{C}}-CH{=}CHR'$$

$$R-\overset{O}{\overset{\|}{C}}-\ddot{C}-CH_2R' \longrightarrow O{=}C{=}\underset{R}{\underset{|}{C}}-CH_2R' \xrightarrow{H_2O} O{=}\underset{HO}{\underset{|}{C}}-\underset{R}{\underset{|}{CH}}-CH_2R'$$

Ethyl trichloroacetyldiazoacetate is another α-diazo ketone that does not undergo the Wolff rearrangement on decomposition (under the conditions studied, at least). Weygand and Koch found that instead a derivative of dichloromaleic acid is formed, perhaps via the opening of a three-membered ring formed by an internal-insertion reaction (199), as shown below.

$$Cl_3C{-}\overset{\overset{\displaystyle O}{\|}}{C}{-}\underset{\underset{\displaystyle N_2}{\|}}{C}{-}CO_2Et \xrightarrow{h\nu} Cl_3C{-}\overset{\overset{\displaystyle O}{\|}}{C}{-}\ddot{C}{-}CO_2Et \rightarrow \overset{\overset{\displaystyle O}{\|}}{C}{-}\overset{\overset{\displaystyle Cl}{|}}{C}{-}CO_2Et \;(\text{ring closed through } CCl_2)$$

$$\downarrow$$

$$\underset{\underset{\displaystyle Cl}{|}}{O{=}C}{-}C(Cl){=}C(Cl){-}CO_2Et \leftarrow O{=}\dot{C}{-}\overset{\overset{\displaystyle Cl}{|}}{C}(Cl){-}\dot{C}(Cl){-}CO_2Et$$

This behavior is not general for acyldiazoacetic esters, several of which have been found by Horner and Spietschka to give the Wolff rearrangement on photolysis (192). In view of the fact that electron-withdrawing substituents usually decrease the migratory aptitude of alkyl groups, it is perhaps not surprising that the Wolff rearrangement is not observed with ethyl trichloroacetyldiazoacetate nor with ethyl trifluoroacetyldiazoacetate, which, in all the cases studied, has been found to give upon photolysis the products of attack on the solvent (171).

Other Rearrangements of Methylenes. In their studies of the generation of methylenes via the decomposition of salts of tosylhydrazones of aliphatic aldehydes and ketones in aprotic solvents, Friedman and Shechter found that β-hydrogen atoms migrate more readily than β-alkyl groups (157). Thus the tosylhydrazone of methyl ethyl ketone gave 68 per cent *trans*-2-butene, 28 per cent *cis*-2-butene, 5 per cent 1-butene, and 0.5 per cent of the internal-insertion product methylcyclopropane, but no isobutylene.

$$CH_3-CH_2-C(=NNTs^-)-CH_3 \xrightarrow{-Ts^-} CH_3-CH_2-\ddot{C}-CH_3 \rightarrow CH_3-CH=CH-CH_3$$

$$CH_3-CH_2-\ddot{C}-CH_3 \downarrow CH_3-CH_3-CH_2=CH_2$$

$$CH_3-CH_2-\ddot{C}-CH_3 \searrow \text{CH}_2\text{—CH}_2 \text{ (cyclopropane ring closed by CH)}, \; CH-CH_3$$

Monoalkylmethylenes, which are believed to be less stable than dialkylmethylenes, appear to be less selective in their reactions; they give more cyclopropanes and when no migration of β-hydrogen can occur there is little rearrangement.

$$CH_3-CH_2-CH=NNTs^- \rightarrow 90\% \; CH_3-CH=CH_2 + 10\% \; \text{cyclo-}(CH_2-CH_2-CH_2)$$

$$CH_3-C(CH_3)_2-CH=NNTs^- \rightarrow 92\% \; \text{cyclo-}(CH_3)_2C\!-\!CH_2\!-\!CH_2 + 7\% \; (CH_3)_2C=CH-CH_3$$

With the tosylhydrazones of certain small-ring compounds, however, there is extensive carbon-skeleton rearrangement.

$$\text{cyclo-}(CH_2-CH_2-CH)-CH=NNTs^- \rightarrow \text{cyclo-}(CH_2-CH_2-CH)-\ddot{C}-H \rightarrow 64\% \; \text{cyclo-}(CH_2-CH_2-CH=CH)$$

$$\text{cyclo-}(CH_2-CH_2-CH)-\ddot{C}-H \downarrow 12\% \; CH_2=CH_2 + CH\equiv CH$$

$$\text{cyclo-}(CH_2-CH_2-CH)-\ddot{C}-H \not\rightarrow \text{cyclo-}(CH_2-CH_2-C)=CH_2$$

Although no methylenecyclopropane was detected in the reaction above it was found to be produced in 80 per cent yield in the decomposition of cyclobutanone tosylhydrazone. Only 19 per cent cyclobutene was formed in this case whereas with cyclopentanone 94 per cent cyclopentene was formed. Yields of cycloalkenes fell, for the medium-ring compounds, to about 20 per cent for cyclononene and cyclodecene, because of competing internal insertion to give bicyclic hydrocarbons (157).

An example of the migration of a fluorine atom has been observed by Fields and Haszeldine, who found that the gas-phase photolysis of trifluorodiazoethane yields about 30 per cent trifluoroethylene, in addition to about 50 per cent 1,1,1,-4,4,4-hexafluoro-2-butene (200).

$$CF_3\text{—}CHN_2 \rightarrow CF_3\text{—}\ddot{C}\text{—}H \rightarrow CF_2{=}CHF$$

$$\downarrow CF_3\text{—}CHN_2$$

$$CF_3\text{—}CH{=}CH\text{—}CF_3$$

Heptafluoropropylmethylene was found to rearrange by migration of a pentafluoroethyl group.

$$C_2F_5\text{—}CF_2\text{—}CHN_2 \rightarrow C_2F_5CF_2\text{—}\ddot{C}\text{—}H \rightarrow CF_2{=}CHC_2F_5$$

$$\downarrow C_3F_7CHN_2$$

$$n\text{-}C_3F_7\text{—}CH{=}CH\text{—}C_3F_7\text{-}n$$

REACTIONS OF METHYLENES WITH DIAZO COMPOUNDS AND WITH EACH OTHER

A number of decomposition reactions of diazo compounds lead to the formation of the products that would be expected from the dimerization of the intermediate methylenes. It appears, however, that in no case is there really good evidence that the "dimer" is actually formed by dimerization of the methylenes. In fact, there is good evidence in several cases that the "dimer" is formed by reaction of the methylene with the diazo compound. In most cases, though, the actual mechanism cannot be ascertained from the available evidence.

Staudinger and Goldstein found that a number of diaryldiazomethanes upon heating, usually in benzene solution, give the corresponding ketazines. Presumably diarylmethylenes are formed and combine with unreacted diaryldiazomethane.

$$Ar_2C{=}N_2 \xrightarrow{-N_2} Ar{-}C{-}Ar \xrightarrow{Ar_2C=N_2} Ar_2C{=}N{-}N{=}CAr_2$$

In carbon disulfide solution diphenyldiazomethane and di-*p*-tolyldiazomethane decompose to give the corresponding tetraarylethylenes. Since the decomposition rate is much faster than in benzene, however, the reaction probably does not consist simply of the formation and dimerization of diarylmethylenes, a reaction that should proceed at about the same rate in carbon disulfide as in benzene. Perhaps a free-radical chain process occurs in carbon disulfide.

The product of the thermal decomposition of diazoisatin has the structure that would result from the dimerization of an intermediate methylene, but one cannot rule out the possibility that a methylene, when formed, reacts with the relatively abundant diazo compounds rather than awaiting the much rarer coincidence of collision with another methylene.

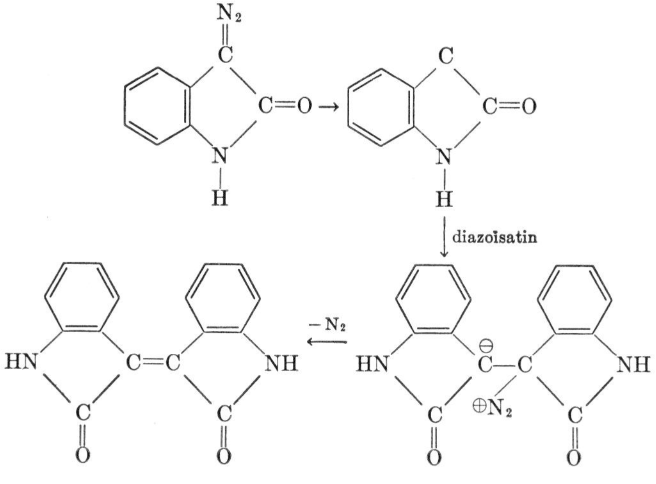

Grundmann found that the copper oxide-catalyzed decomposition of diazoacetophenone leads to 1,2-dibenzoylethylene and the thermal decomposition leads to 1,2,3-tribenzoylcyclopropane. It is reasonable to assume that one molecule of methylene, but probably no more, is involved in the formation of the latter compound.

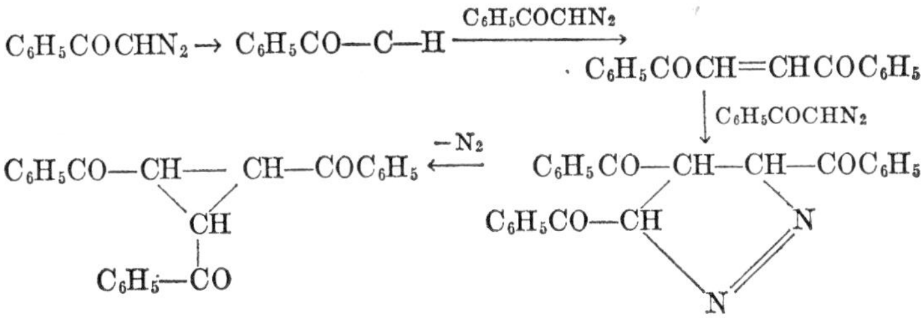

$$C_6H_5COCHN_2 \rightarrow C_6H_5CO\text{—}C\text{—}H \xrightarrow{C_6H_5COCHN_2} C_6H_5COCH{=}CHCOC_6H_5$$

$$C_6H_5COCH{=}CHCOC_6H_5 \xrightarrow{C_6H_5COCHN_2} \underset{\text{(ring: CH—N=N—)}}{C_6H_5CO\text{—}CH\text{—}CH(COC_6H_5)\text{—}CH(COC_6H_5)}$$

$$C_6H_5CO\text{—}CH\text{—}CH\text{—}COC_6H_5 \ (\text{cyclopropane ring with } CH\text{—}COC_6H_5) \xleftarrow{-N_2} \text{pyrazoline}$$

The copper oxide-catalyzed reactions of diazo ketones, which probably do not involve methylene intermediates, have been studied further by Ernest and coworkers.

Kirmse and Horner observed that the photolysis of 4,5-diphenyl-1,2,3-thiodiazole (thiodiazoles exist in the form of heterocyclic compounds rather than as α-diazo thioketones) yields tetraphenyl-1,4-dithiacyclohexadiene (XV), which can be written as arising from the dimerization of an intermediate methylene but which is probably formed by attack of a methylene on unreacted thiodiazole, as shown below.

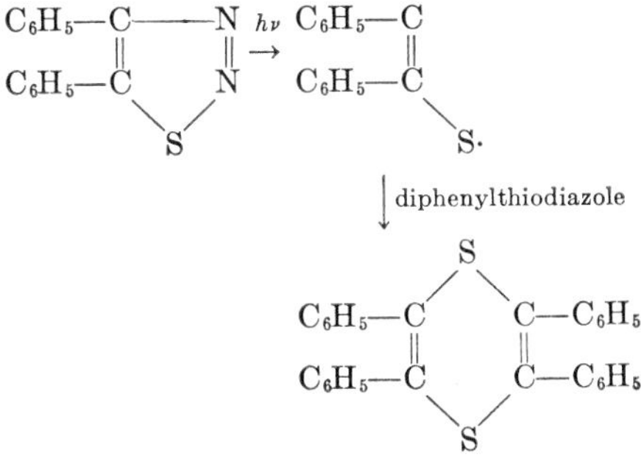

$$\begin{array}{c} C_6H_5\text{—}C\text{——}N \\ \| \qquad \| \\ C_6H_5\text{—}C \quad N \\ \diagdown \ \diagup \\ S \end{array} \xrightarrow{h\nu} \begin{array}{c} C_6H_5\text{—}\dot{C} \\ \| \\ C_6H_5\text{—}C \\ \diagdown \\ S\cdot \end{array}$$

$$\big\downarrow \text{diphenylthiodiazole}$$

$$\begin{array}{c} S \\ \diagup \quad \diagdown \\ C_6H_5\text{—}C \qquad C\text{—}C_6H_5 \\ \| \qquad\qquad \| \\ C_6H_5\text{—}C \qquad C\text{—}C_6H_5 \\ \diagdown \quad \diagup \\ S \end{array}$$

XV

The photolysis of 4-phenyl-1,2,3-thiodiazole, like that of the other monosubstituted thiodiazoles studied, gave no dithiacyclohexadiene but only the 1,4-dithiafulvene derivative (XVII) shown below, which must have resulted from addition of the intermediate methylene to the thioketene XVI (201). Dithiafulvenes are by-products in the photolysis of disubstituted thiodiazoles.

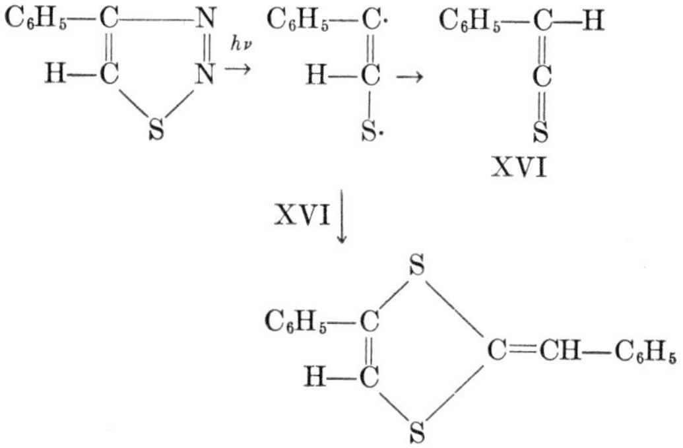

$$C_6H_5\text{—}C{=}C(H)\text{—}S\text{—}N{=}N \text{ (ring)} \xrightarrow{h\nu} C_6H_5\text{—}\dot{C}{=}CH\text{—}\dot{S} \rightarrow C_6H_5\text{—}CH{=}C{=}S \;(\text{XVI})$$

$$C_6H_5\text{—}\dot{C}{=}CH\text{—}\dot{S} \xrightarrow{\text{XVI}} \text{XVII}$$

XVII: ring C₆H₅—C=C(H) bridged by two S atoms to C=CH—C₆H₅

Whether the intermediate methylenes are singlets, or triplets as implied in the equations above, is not known.

DeMore, Pritchard, and Davidson found that the interesting hydrocarbon fulvalene (XVIII) is produced by the photolysis of diazocyclopentadiene in a rigid nitrogen matrix at 20°K or a fluorocarbon matrix at 77° K, but not in a hydrocarbon matrix at 77° K (202).

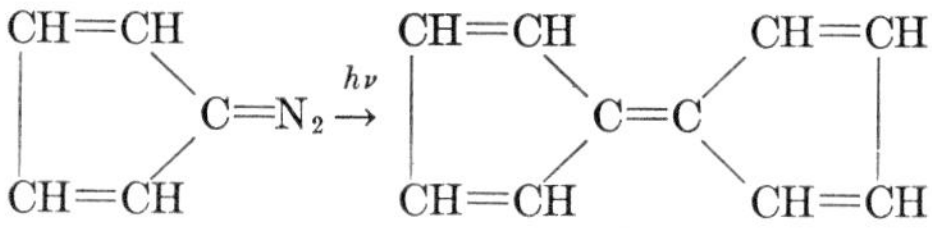

$$\text{(CH=CH)}_2\text{C=N}_2 \xrightarrow{h\nu} \text{(CH=CH)}_2\text{C=C(CH=CH)}_2 \;(\text{XVIII})$$

Whether the fulvalene is produced by dimerization of the methylene or by attack of the methylene on the diazo compound is an open question.

The photolysis of diazoethane was found by Brinton and Volman to yield 2-butene, acetylene, and hydrogen as well as ethylene. The formation of 2-butene is attributable to the attack of ethylidene on diazoethane and it provides evidence that the major product ethylene is not formed by a concerted reaction but via the intermediate ethylidene. Frey studied the reaction in detail and was able to explain his observations in terms of the reaction mechanism below, according to which the photolysis initially gives ethylidene, which may rearrange to an excited molecule (containing much excess energy) of ethylene or react with diazoethane to give an excited molecule of *trans*-2-butene. The excited ethylene may either give up its excess energy by collision with some molecule (M in the scheme below) or decompose to acetylene and hydrogen. The excited *trans*-2-butene may either isomerize (reversibly) to excited *cis*-2-butene or become deactivated (203).

$$CH_3CHN_2 \xrightarrow{h\nu} CH_3{-}C{-}H + N_2$$
$$CH_3{-}C{-}H \rightarrow C_2H_4^*$$
$$C_2H_4^* \rightarrow C_2H_2 + H_2$$
$$C_2H_4^* + M \rightarrow C_2H_4 + M$$
$$CH_3{-}C{-}H + CH_3CHN_2 \rightarrow trans\text{-}CH_3CH{=}CHCH_3^* + N_2$$
$$trans\text{-}CH_3CH{=}CHCH_3^* \rightleftharpoons cis\text{-}CH_3CH{=}CHCH_3^*$$
$$cis\text{- or } trans\text{-}CH_3CH{=}CHCH_3^* + M \rightarrow cis\text{- or } trans\text{-}CH_3CH{=}CHCH_3 + M$$

In agreement with this mechanism the yield of 2-butenes increases at higher pressures of diazoethane, where the intermediate ethylidene molecules more often encounter a diazoethane molecule before they isomerize to excited ethylene. The yield of acetylene and the fraction of the 2-butene that is *cis* both increase with decreasing pressure, since at lower pressures the excited ethylene has more time to lose hydrogen and the excited *trans*-2-butene has more time to isomerize before being deactivated. The agreement is quantitative as well as qualitative.

In the presence of propane and butane no products of insertion by ethylidene were observed. With propylene a small amount of insertion may have occurred but the principal reaction was addition, leading to a mixture of *cis*- and *trans*-1,2-dimethylcyclopropane.

$$CH_3—C—H + \underset{\underset{CH_2}{\|}}{CH}—CH_3 \rightarrow \textit{cis- and trans-}\ CH_3—\underset{\diagdown\ \diagup}{CH——CH}—CH_3 \quad (\text{ring } CH_2)$$

Propylene is only about one-twentieth as reactive as diazoethane is toward ethylidene, and both 2-butenes were found to be considerably less reactive than propylene. This latter observation may seem surprising in view of the fact that the 2-butenes are more reactive than 1-alkenes toward dihalomethylenes and carbethoxymethylene. It may suggest that ethylidene is a largely nucleophilic methylene unlike the largely electrophilic carbalkoxy- and dihalo-methylenes. However, even in the absence of polar effects, there are several factors that could make the 2-butenes less reactive than propylene toward ethylidene. First, although steric hindrance should not be large it should favor reaction with propylene. Second, judging from heats of hydrogenation, *cis*- and *trans*-2-butene are 1.5 and 2.5 kcal./mole more stable, respectively (relative to the saturated hydrocarbon), than propylene. The differences in stabilities of the cyclopropane derivatives being formed are probably qualitatively similar but quantitatively much smaller. Therefore the addition of ethylidene to propylene should be more favorable energetically than its addition to the 2-butenes. To whatever extent this difference in energies of reaction is reflected in energies of activation the additions to the butenes will be slower. Hence the relative reactivities of propylene and the 2-butenes toward ethylidene do not show whether ethylidene behaves largely as a nucleophilic or an electrophilic reagent in its reactions with olefins; the data merely show that the electrophilicity of ethylidene is not so great as to override the other factors influencing the relative reaction rates.

THE DECOMPOSITION OF 1,4-BIS(α-DIAZOBENZYL)BENZENE

Murray and Trozzolo have studied the decomposition of 1,4-bis(α-diazobenzyl)benzene (XIX), a compound of particular interest because of the possibility that it yields an intermediate in which there are two divalent carbon atoms. Just as the photolysis of diphenyldiazomethane in the presence of oxygen is known to yield benzophenone, the photolysis of XIX yields 1,4-dibenzoylbenzene. The partial photolytic decomposition of XIX in the presence of oxygen gives a mixture of dibenzoylbenzene and unchanged XIX, but no diazoketone was observed. This observation suggests that the reaction proceeds by the intermediate formation of XX, which has been shown by EPR studies to have only two unpaired electrons (204).

$$C_6H_5-C(=N_2)-C_6H_4-C(=N_2)-C_6H_5 \ (\text{XIX}) \xrightarrow{h\nu} \left[C_6H_5-\dot{C}=C_6H_4=\dot{C}-C_6H_5 \leftrightarrow \text{etc.} \right] (\text{XX}) \xrightarrow{O_2} C_6H_5-C(=O)-C_6H_4-C(=O)-C_6H_5$$

Crystalline XIX displayed the interesting property of pleochroism; the absorption spectrum of the crystals varied with the plane of polarization of the incident light. When viewed with polarized light under a microscope some of the crystals were purple and others colorless. When the microscope stage was rotated the colorless crystals became purple and vice versa. When the light was intense the purple crystals decomposed giving off bubbles of nitrogen, but the colorless crystals were not affected.

8

Miscellaneous Methods of Forming Substituted Methylenes

DECOMPOSITION OF KETENES

Kistiakowsky and Mahan found that the gas-phase photolysis of methyl ketene yields carbon monoxide, ethylene, and 2-butene (205). The yield of carbon monoxide is equal to the yield of ethylene plus twice the yield of butene. The ratio of butene to ethylene formed increases with increasing pressure. These observations are in agreement with the mechanism

$$CH_3CH{=}CO \rightarrow CH_3{-}\ddot{C}{-}H + CO$$

$$CH_3{-}\ddot{C}{-}H \swarrow CH_2{=}CH_2$$

$$CH_3{-}\ddot{C}{-}H \xrightarrow{CH_3CH=CO} CH_3CH{=}CHCH_3 + CO$$

according to which the rate of attack of ethylidene on methyl ketene is comparable to its rate of rearrangement under the conditions used. Under a given set of conditions the ethylene yield decreases when an inert gas is added and when light of a longer wave length is used. It therefore appears that the presence of excess energy in the intermediate ethylidene molecules increases their isomerization rate more than it increases their rate of attack on methyl ketene.

The photolysis of dimethyl-ketene vapor was found by Holroyd and Blacet to yield largely carbon monoxide and propylene (206). Rearrangement of the intermediate methylene to an olefin occurs even faster (relative to attack of the methylene on ketene) in this case than in the decomposition of methyl ketene. The quantum yield at 2537 Å is 1.06 ± 0.06, and little, if any, hexene is formed.

α-DEHALOGENATION

Dicyanomethylene. Dicyanomethylene may be written as an intermediate in the preparation of tetracyanoethylene from dibromomalononitrile and copper powder in boiling benzene (207), as well as in a number of other reactions. In support of such a reaction mechanism Cairns and coworkers found that when the reaction is carried out in cyclohexene instead of benzene, cyclohexylidenemalononitrile can be isolated in 15 per cent yield. It was proposed that the cyclohexylidenemalononitrile arose from the isomerization of initially formed 7,7-dicyanobicyclo[4.1.0]heptane (I), as shown below.

$$N{\equiv}C{-}CBr_2{-}C{\equiv}N \xrightarrow{Cu} \left[N{\equiv}C{-}\dot{C}{\cdot}{-}C{\equiv}N \leftrightarrow \cdot N{=}C{=}C{=}C{=}N\cdot \leftrightarrow N{\equiv}C{-}C\cdot{=}C{=}N\cdot \leftrightarrow \cdot N{=}C{=}C\cdot{-}C{\equiv}N \leftrightarrow \cdot N{-}C{\equiv}C{-}C{\equiv}N \leftrightarrow \text{etc.} \right]$$

$$\xrightarrow{\text{cyclohexene}} \text{7,7-dicyanobicyclo[4.1.0]heptane (I)} \rightarrow C_6H_{10}{=}C(CN)_2$$

When ethyl dibromocyanoacetate was used instead of dibromomalononitrile the bicyclic cyanoester II was isolated in 10 per cent yield.

$$NC{-}CBr_2{-}CO_2Et \xrightarrow[\text{cyclohexene}]{Cu} C_6H_{10}{>}C(CN)(CO_2Et) \quad \text{II}$$

As shown in the reaction scheme leading from dibromomalononitrile, the triplet form of dicyanomethylene should be greatly stabilized by resonance; in the molecular-orbital representation of this intermediate the five atoms are arranged in a straight line and surrounded by a cylindrically symmetrical sheath of ten electrons.

Dicyanomethylene is thus a very attractive possibility as an intermediate in the reaction of dibromomalononitrile with copper. However, in view of the evidence that organometallic compounds, including organocopper compounds, may also react with olefins to yield cyclopropane derivatives, it is difficult to rule out certain alternative mechanisms for the reaction. Some of these alternatives and the dicyanomethylene-dimerization mechanism suggested by Cairns and coworkers, are shown below.

$$CBr_2(CN)_2 \xrightarrow{Cu} Cu{-}C(CN)_2{-}Br \rightarrow :C(CN)_2$$

$$CBr_2(CN)_2 \rightarrow Br{-}\dot{C}(CN)_2 \nearrow Cu{-}C(CN)_2{-}Br$$

$$Cu{-}C(CN)_2{-}Br \xrightarrow{Br_2C(CN)_2} Br{-}C(CN)_2{-}C(CN)_2{-}Br$$

$$2\,Br{-}\dot{C}(CN)_2 \xrightarrow{\text{dimerization}} Br{-}C(CN)_2{-}C(CN)_2{-}Br \xrightarrow{Cu} (NC)_2C{=}C(CN)_2$$

$$2\,:C(CN)_2 \xrightarrow{\text{dimerization}} (NC)_2C{=}C(CN)_2$$

Cyclopropylidene and Its Derivatives. It seems that 1,1-dihalocyclopropanes are surrounded by methylenes. These compounds were hardly more than laboratory curiosities until their preparation by the generation of dihalomethylenes in the presence of olefins was discovered; only a few years later it was discovered that 1,1-dihalocyclopropanes themselves may be used to generate methylenes of a very interesting type. Doering and LaFlamme found that the reaction of 1,1-dibromocyclopropane derivatives with sodium or magnesium gives allenes (208).

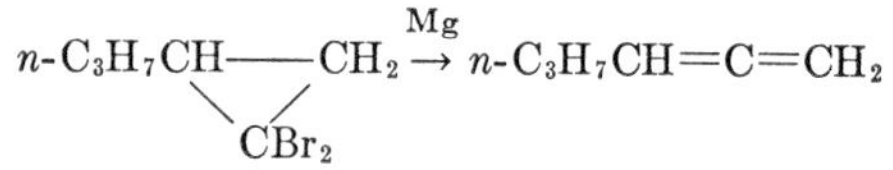

$$n\text{-}C_3H_7CH\overset{\displaystyle CH_2}{\underset{\displaystyle CBr_2}{\triangle}} \xrightarrow{Mg} n\text{-}C_3H_7CH{=}C{=}CH_2$$

They pointed out some of the arguments against free-radical and carbanion mechanisms for the ring-opening reactions and suggested the intermediate formation of a derivative of cyclopropylidene, which then undergoes ring opening to give the allene directly.

Alkyllithium compounds may also be used to bring about the transformation of 1,1-dihalocyclopropanes to allenes. Skattebøl described a number of examples of this type of reaction, including the introduction of two sets of allenic double bonds into a ten-membered ring (209).

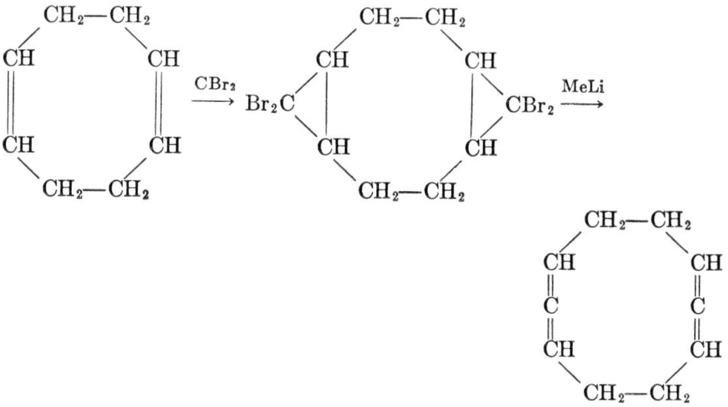

$$\text{cyclo-}(CH_2CH_2CH{=}CHCH_2CH_2CH{=}CH) \xrightarrow{CBr_2} \text{bis}(CBr_2\text{-cyclopropane adduct}) \xrightarrow{MeLi} \text{cyclo-}(CH_2CH_2CH{=}C{=}CHCH_2CH_2CH{=}C{=}CH)$$

Moore, Ward, and Merritt found that the reaction of 7,7-dibromobicyclo[4.1.0]heptane with methyllithium in ether gives several unusual products, whose formation can be explained in terms of the reactions of an intermediate cyclopropylidene (210). The highly strained bicyclic compounds IV and V presumably arose from internal insertion, and product VI from dimerization by the intermediate cyclopropylidene III. Insertive attack on the solvent would explain the formation of VII.

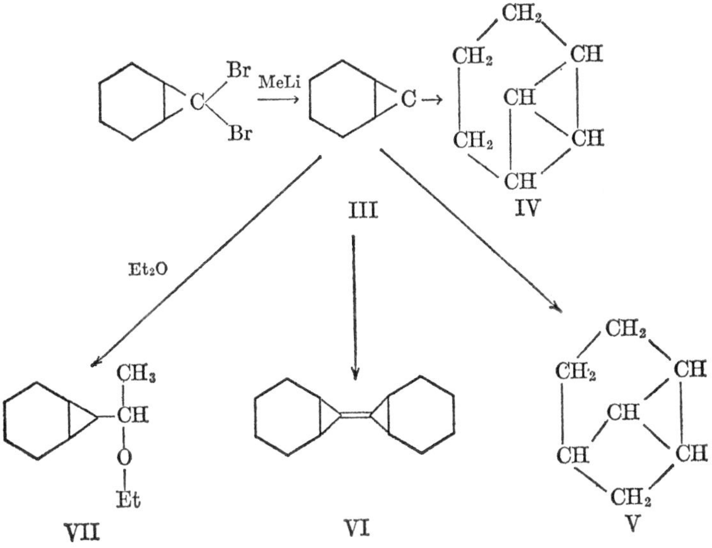

MeLi

III

IV

Et_2O

VII

VI

V

In the presence of olefins, derivatives of spiropentane are formed.

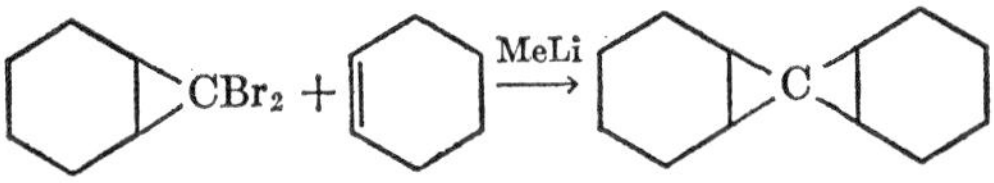

CBr_2 + MeLi

Utilization of *cis*- and *trans*-2-butene shows that the addition to olefins is stereospecific.

Although addition to olefins and both internal and external insertion have been observed for the bicyclic cyclopropylidene III, no such reactions have been observed for cyclopropylidenes generated from monocyclic 1,1-dihalocyclopropanes. These facts are reasonably explained by the large amount of energy that might be required to bring about the rearrangement of III to the highly strained molecule 1,2-cycloheptadiene. Since III cannot so readily rearrange to an allene its lifetime is long enough so that other types of reactions can be observed. It may alternatively be suggested that the formation of allenes from monocyclic dihalocyclopropanes occurs so readily that it is a partially or completely concerted reaction, not involving the formation of a cyclopropylidene. This explanation seems less likely, however, especially in view of the fact that a cyclopropylidene appears to be formed (spiropentane derivatives are obtained in the presence of olefins) when the N-nitrosourea from 2,2-diphenylcyclopropylamine is decomposed by base (cf. p. 133).

Miscellaneous α-Dehalogenations. Yellow luminescence was found by Bawn and Dunning to be produced in the vapor-phase reaction of sodium with organic dihalides at 300° and about 10^{-3} mm. pressure (211). Organic monohalides, under the same conditions, react without producing luminescence. From the latter observation it appears that the reactions

$$RX + Na \rightarrow R\cdot + NaX$$

and

$$2R\cdot \rightarrow R{-}R \text{ (or disproportionation products)}$$

are not sufficiently exothermic to excite the *D* line of sodium. It therefore appears that the reaction of the first atom of sodium with an organic dihalide gives an intermediate (perhaps with considerable excess energy) whose reaction with a second sodium atom yields enough energy to cause luminescence, e.g.,

$$X{-}R{-}X + Na \rightarrow X{-}R\cdot + NaX$$
$$X{-}R\cdot + Na \rightarrow \cdot R\cdot + NaX^*$$
$$NaX^* + Na \rightarrow Na^* + NaX$$

The second step of the reaction particularly should be highly exothermic when the dihalide used is one like ethylene bromide or propylene bromide, since in such cases two stable molecules are formed. The reaction of ethylidene bromide was accompanied by much less luminescence than that of ethylene bromide, as would be expected, since an α-bromoethyl radical should react with sodium much less exothermically than a β-bromoethyl radical.

$$CH_3\dot{C}HBr + Na \rightarrow NaBr + CH_3{-}C{-}H$$
$$\cdot CH_2CH_2Br + Na \rightarrow NaBr + CH_2{=}CH_2$$

In fact, Bawn and Milsted thought that the first of the two reactions above could not give off enough energy to result in luminescence. For this reason and the fact that no dimerization of ethylidene to butene was observed and no ethane was found when the reaction was run in the presence of hydrogen, it was concluded that the second step in the reaction of ethylidene with sodium yields ethylene (the observed product) directly.

$$CH_3\dot{C}HBr + Na \rightarrow NaBr + CH_2{=}CH_2$$

In view of our present knowledge of ethylidene, however, it seems probable that dimerization and reaction with hydrogen would probably not have been observable in the studies of Bawn and coworkers, and that ethylidene may very well have been a real intermediate in the reaction of ethylidene bromide with sodium.

The generation of methylenes by the reaction of *gem*-dibromides with organolithium compounds is not a unique reaction of cyclopropane derivatives; it appears to occur rather generally. Kirmse, for example, reported that the reaction of methyllithium with 2,2-dibromo-3,3-dimethylbutane yields 1,1,2-trimethylcyclopropane and *t*-butylethylene (128).

$$\begin{array}{c}CH_3\\|\\CH_3-C-CH_3\\|\\CBr_2\\|\\CH_3\end{array}\xrightarrow{MeLi}\begin{array}{c}CH_3\\|\\CH_3-C-CH_3\\|\\C\\|\\CH_3\end{array}\rightarrow\begin{array}{c}CH_3\\|\\CH_3-C-CH_2\\|\ /\\CH\\|\\CH_3\end{array}$$

$$\downarrow$$

$$\begin{array}{c}CH_3\\|\\CH_3-C-CH_3\\|\\CH\\||\\CH_2\end{array}$$

Reaction of methyllithium with 2,2-dibromobutane yields 2 per cent 1-butene, 24 per cent *cis*-2-butene, and 74 per cent *trans*-2-butene, a product mixture with almost the same composition as that produced by the decomposition of 2-diazobutane. It should not be forgotten, however, that the intermediate produced by the reaction of diphenyldibromomethane with alkyllithium behaves differently from the one produced by the decomposition of diphenyldiazomethane (cf. p. 128).

DISSOCIATION OF TETRASUBSTITUTED ETHYLENES TO METHYLENES

Dissociation of Tetraarylethylenes. Just as hexaarylethanes are known to dissociate readily under mild conditions, it might be expected that substituted ethylenes would dissociate to methylenes if the substituents were such as to provide sufficient stabilization for the methylene (relative to the olefin). With the thought that aryl groups might provide such stabilization Staudinger and Goldstein studied the decomposition of bis(*p*-phenylphenyl)diazomethane and several other diaryldiazomethanes, hoping to be able to isolate a stable diarylmethylene. Although no stable diarylmethylene was found by these or by subsequent workers, Franzen and Joschek suggested that the decomposition of tetra-α-naphthylethylene at temperatures around 200° is initiated by a cleavage to di-α-naphthylmethylene (212). Their evidence consists chiefly of the observation that tetra-α-naphthylethylene in benzylamine at 200° yields some N-(di-α-naphthyl-methyl)benzylamine in addition to the major product 13*H*-dibenzo[*a*,*g*]fluorene.

$$(\alpha\text{-}C_{10}H_7)_2C{=}C(C_{10}H_7\text{-}\alpha)_2 \rightarrow \alpha\text{-}C_{10}H_7{-}\ddot{C}{-}C_{10}H_7\text{-}\alpha$$

$$\xrightarrow{C_6H_5CH_2NH_2} (\alpha\text{-}C_{10}H_7)_2CHNHCH_2C_6H_5$$

(dibenzofluorene structure with CH_2)

The fact that tetra-α-naphthylethylene decomposes under conditions where tetraphenylethylene is stable may be explained in part by the instability due to steric interactions in the former olefin. Such steric interactions should be negligible in the diarylmethylene, where the two aryl groups presumably lie in perpendicular planes. In addition, di-α-naphthylmethylene must be stabilized more by resonance than is diphenylmethylene.

Bis-(1,3-diphenyl-2-imidazolidinylidene). Wanzlick and coworkers described evidence that a particularly stable methylene may be an intermediate in the formation and reactions of bis-(1,3-diphenyl-2-imidazolidinylidene) (compound VIII) (213). Compound VIII may be prepared from 1,2-dianilinoethane by heating the imidazolidine IX formed by reaction with chloral, or simply by heating with orthoformic ester. It was suggested that both reactions involve the intermediate formation of the methylene X (by an unstated mechanism). (See next page.)

Although the molecular weight calculated from the formula of VIII is 445, three Rast molecular-weight determinations

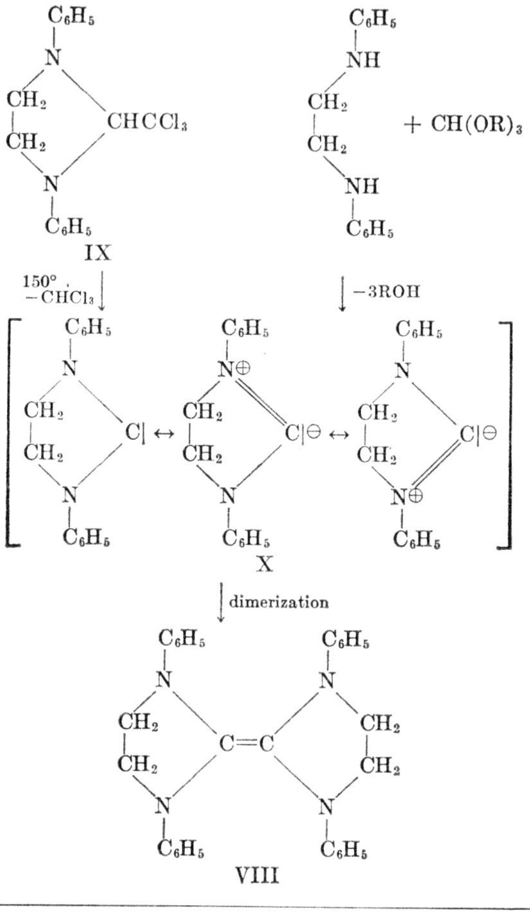

$$\underset{\text{IX}}{\text{1,3-diphenyl-2-(trichloromethyl)imidazolidine: } C_6H_5N(CH_2CH_2)NC_6H_5 \text{ ring with } CHCCl_3} \xrightarrow[-CHCl_3]{150^\circ} \left[\text{carbene} \leftrightarrow N^{\oplus}{=}C|^{\ominus} \leftrightarrow C|^{\ominus}{=}N^{\oplus}\right] \ (\text{X}) \xleftarrow{-3ROH} C_6H_5NHCH_2CH_2NHC_6H_5 + CH(OR)_3$$

$$\text{X} \xrightarrow{\text{dimerization}} \underset{\text{VIII}}{\text{bis(1,3-diphenylimidazolidin-2-ylidene) } C{=}C}$$

in the absence of oxygen gave a molecular weight of 305 ± 12. Since pure unchanged VIII was recovered after the determination, the low molecular weight was attributed to partial (about 60 per cent) dissociation to the methylene X. As shown in the following reaction scheme, VIII was found to react with a variety of reagents to give products whose formation is easily rationalized on the basis of the intermediate formation of the methylene X.

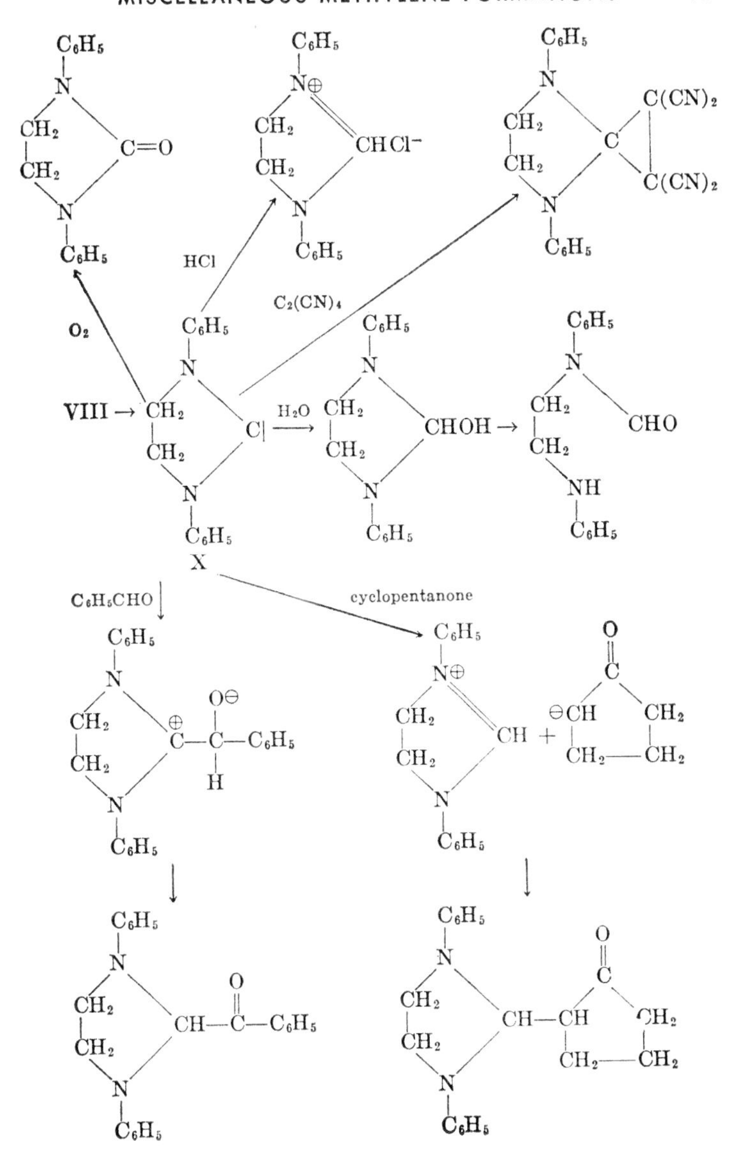

HCl
$C_2(CN)_4$
O_2
VIII →
H_2O
X
C_6H_5CHO
cyclopentanone

At least some of these reactions may also be explained plausibly without invoking a methylene intermediate. Kinetic studies should be capable of distinguishing mechanisms involving direct attack on VIII from those involving the intermediate formation of X. No such studies appear to have been reported yet, however.

In view of the highly electrophilic character of dihalomethylenes and the evidence that such methylenes may be stabilized by sharing the unshared electron pairs of the α-halogen substituents, it seems probable that much of the unusual stability of the methylene X (relative to its dimer VIII) is due to the much greater ability of amino substituents to share their unshared electron pairs. It is also clear, however, that in the dimer VIII there is interference between the phenyl groups attached to the two halves of the molecule; this probably contributes considerably to the ease of dissociation of VIII to X. More about the relative importance of electronic and steric factors might be learned by study of the compound in which the phenyl groups of VIII have been replaced by methyl groups. This would certainly increase the electron-donating power of the α-amino substituents on the divalent carbon atom of the methylene and would probably decrease the steric strain in the methylene dimer.

As Wanzlick pointed out, the electron-donating power of the amino substituents is sufficient to make the methylene X a predominantly nucleophilic reagent; its reactions appear to be initiated by proton acceptance and by attack at such electrophilic centers as carbonyl carbon atoms.

STABLE METHYLENES

Although such stable derivatives of double-bonded divalent carbon as carbon monoxide and isocyanides are known, no report of the preparation of a divalent carbon derivative of the type X—C—Y as a pure compound appears to have won general acceptance.* Preparing such a stable X—C—Y may

* Reports of the preparation of carbon monoxide diethyl acetal (EtO—C—OEt) and dichloromethylene as pure substances are believed to be in error.

be viewed as stabilizing methylene itself by replacing its hydrogen atoms by suitable X and Y groups. Stabilization of the singlet and triplet forms of methylene presents two separate problems.

The singlet form of methylene has an unshared electron pair, like a carbanion, and only six electrons in its outer shell, like a carbonium ion, and is therefore both a nucleophilic and an electrophilic reagent. Since isolation of X—C—Y as a pure compound demands only that it not react with itself, it might be sufficient merely to put in substituents to reduce the nucleophilicity and isolate pure X—C—Y as a highly electrophilic species, or merely to put in substituents to reduce the electrophilicity and isolate pure X—C—Y as a highly nucleophilic species. A salt containing the dianion of formic acid might prove to be a suitable example of the latter-type species. The dianion of formic acid appears to be an intermediate in a number of reactions of formate ions (e.g., deuterium exchange) but so far appears to have been generated only in media containing species with which it reacts. In any event the pure salt might have a carbon-metal bond with considerable covalent character and therefore not be a methylene.

It should be more convenient to isolate a stable methylene that is neither strongly nucleophilic nor strongly electrophilic, having substituents that stabilize both carbanions and carbonium ions. Qualitatively, α-halogen substituents are of this type and this is probably why dihalomethylenes may be generated readily in a number of ways. However, α-halogen substituents are not *highly* effective at either carbonium-ion or carbanion stabilization and for this reason dihalomethylenes are quite reactive species. As α-substituents aryl groups are also capable of stabilizing both carbanions and carbonium ions; they are also effective at stabilizing free radicals, however, and this, no doubt, is partly why the most stable form of diphenylmethylene is apparently a triplet. It might be more effective to use one electron-donating and one electron-withdrawing α-substituent. For this reason attempts have been made to generate *p*-nitrophenyl-*p*-dimethylaminophenylmethylene (XI), which, if a singlet, would be stabilized by resonance as shown below.

NO_2 NO_2 $NO_2^{\ominus}$

C| ↔ C|⊖ ↔ C ↔ etc.

NMe_2 $\oplus NMe_2$ $\oplus NMe_2$

XI

Conclusive evidence for the formation of XI has not been obtained but work is continuing.

The presence of bulky ortho substituents on the aromatic rings of XI should greatly decrease the stability of the dimer, a tetraarylethylene whose aromatic rings should have a tendency to be coplanar. There should be much less effect on the stability of the methylene. One might also consider attaching the electron-donating dimethylamino group and the electron-withdrawing nitro group to the divalent carbon atom directly rather than through a phenyl group as in XI. The resultant methylene would also be highly resonance-stabilized as shown below, but no great steric hindrance to dimerization would be present.

$$\left[\begin{array}{ccccccc} NO_2 & & NO_2^{\ominus} & & NO_2^{\ominus} & & NO_2 \\ | & & \| & & \| & & | \\ \mathrm{C}| & \leftrightarrow & \mathrm{C} & \leftrightarrow & \mathrm{C}^{\oplus} & \leftrightarrow & \mathrm{C}|^{\ominus} \\ | & & \| & & | & & \| \\ NMe_2 & & {}^{\oplus}NMe_2 & & NMe_2 & & {}^{\oplus}NMe_2 \end{array}\right]$$

This methylene also does not appear to have been reported. Greater stability would be expected if the dimethylamino group were replaced by the even more strongly electron-donating alkoxide anion group (—O⁻) and if the nitro group were replaced by the more strongly electron-withdrawing $—CH_2^+$ group. The resultant "methylene"

$$\ominus O—\bar{C}—CH_2\oplus$$

is a known compound that has been isolated in the pure state, but, of course, it is better known when represented by the formula

$$O=C=CH_2$$

and the name ketene.

The preceding exercise in *reductio ad absurdum* illustrates the point that in any consideration of the preparation of stable (or even unstable) derivatives of divalent carbon some criterion for the acceptance of a given carbon atom as divalent is needed. To provide such a criterion it is suggested that X—C—Y be considered a divalent-carbon derivative if there exists as a stable compound some species such as

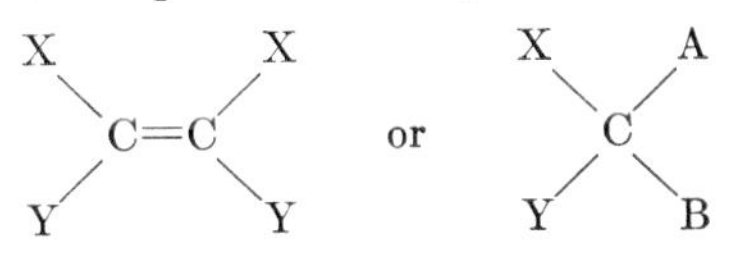

This criterion may not be entirely satisfactory but it does exclude ketene from the methylene category while including X, XI, and all the other species generally considered to contain divalent carbon.

Many methylenes may be thought of as the conjugate bases of formic-acid derivatives. The fact that formate esters are readily decarbonylated by alkoxide ions shows that the anion

$$\left[R—O—\overset{\ominus}{\bar{C}}=O \leftrightarrow R—O—\bar{C}—\overset{\ominus}{\underline{\bar{O}}}| \right]$$

is moderately stable (assuming that the decarbonylation is a stepwise rather than a concerted reaction). Replacement of the =O substituent by the more strongly electron-withdrawing $=NR_2^+$ group transforms a formate ester to the quaternary salt derived from an imino ether.

$$RO—CH=NR_2^+$$
XII

The acidity of XII should be greater than that of a formate ester. Replacement of the alkoxy group of XII by a dialkylamino group, yielding a formamidinium salt, should decrease the acidity of the hydrogen somewhat.

$$\left[\begin{array}{c}\oplus NR_2\\ \| \\ H-C\\ | \\ NR_2\end{array} \leftrightarrow \begin{array}{c}NR_2\\ | \\ H-C\\ \| \\ \oplus NR_2\end{array}\right] \xrightarrow{-H^+} \left[\begin{array}{c}\oplus NR_2\\ \| \\ \ominus|C\\ | \\ NR_2\end{array} \leftrightarrow \begin{array}{c}NR_2\\ | \\ \ominus|C\\ \| \\ \oplus NR_2\end{array} \leftrightarrow \text{etc.}\right]$$

The acidity of amidinium salts may still be greater than that of formate esters, though; Wanzlick's methylene (X), which is the conjugate base of a formamidinium cation, appears to be moderately stable.

Since α-alkylthio substituents are known to stabilize carbanions, the replacement of the alkoxy group of XII by an alkylthio group should yield an even more acidic species. One example of the resultant type of species is, as Breslow pointed out, thiamine (vitamin B_1). Breslow found that thiamine and other N-substituted thiazolium salts undergo deuterium exchange readily in neutral aqueous solution at room temperature in the absence of catalysts (214). This deuterium exchange was shown to occur at the carbon atom between nitrogen and sulfur and it must involve the intermediate formation of species XIII.

$$\left[\;\text{R–C}=\text{C–R ring: } \overset{R}{N^{\oplus}}=C|^{\ominus},\ S \;\leftrightarrow\; \overset{R}{N|}-C|-S \;\leftrightarrow\; \overset{R}{N^{\oplus}}=C=S^{\ominus} \;\leftrightarrow \text{etc.}\right]$$

XIII

To whatever extent the middle of the three valence-bond structures written above contributes to the total structure, XIII may be considered a methylene. Breslow and others have presented convincing evidence that the formation of

XIII is of fundamental importance in the biological action of thiamine.

The stabilization of triplet methylene would seem to be best accomplished by introducing groups to permit the unpaired electrons to spend appreciable amounts of time in a large number of different atoms and by sterically inhibiting dimerization. Both of these objectives have been partly accomplished in diphenylmethylene, which nevertheless appears to be capable of abstracting hydrogen atoms from saturated hydrocarbons (162, 163). This reactivity of diphenylmethylene is probably due in part to the fact that a highly resonance-stabilized benzhydryl radical is formed in the reaction. Di-α-naphthylmethylene appears to be more easily formed than diphenylmethylene, presumably because dimerization is hindered more and because the α-naphthyl group is a better radical-stabilizing substituent than the phenyl group (212). Di-9-anthrylmethylene should be much more stable than di-α-naphthylmethylene.

Zimmerman and Paskovich found that such ortho substituents as methyl and chlorine, when present in diphenyldiazomethanes, are sufficient to prevent hydrogen abstraction from the solvent. However, dimerization to tetraarylethylenes does occur and with dimesitylmethylene internal insertion is also observed.

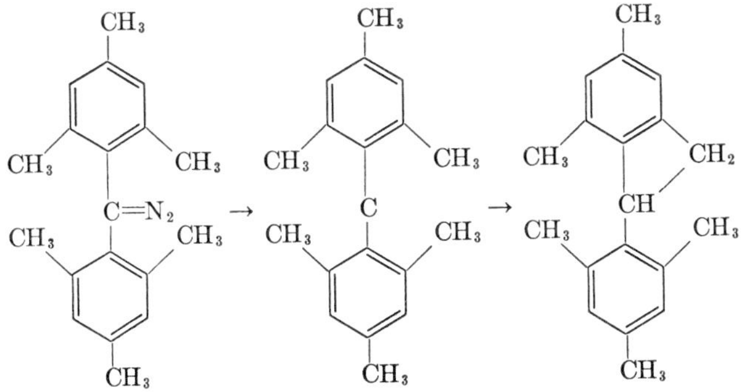

CH_3 CH_3 CH_3

CH_3 CH_3 CH_3 CH_3 CH_3 CH_2

$C{=}N_2$ $\rightarrow$ C $\rightarrow$ CH

CH_3 CH_3 CH_3 CH_3 CH_3 CH_3

CH_3 CH_3 CH_3

Continuation of studies on such species as dicyanomethylene, propargylene, etc., may yield triplet methylenes whose free-radical reactivity is greatly diminished by resonance stabilization. Such studies may be made quite difficult, however, by the highly unstable character of some of the required polyacetylene precursors.

FORMATION OF ALKYLMETHYLENES FROM HALOMETHYLENES AND ORGANOMETALLIC COMPOUNDS

Evidence for the formation of an alkenylmethylene by the reaction of an alkenyllithium with chloromethylene was described in the previous chapter. Closs has obtained stronger evidence that alkyllithiums react with chloromethylene to give alkylmethylenes. The reaction of *n*-butyllithium with methylene chloride had previously been shown to yield 1-pentene. However it seemed possible that the 1-pentene was formed by the dehydrochlorination of *n*-pentyl chloride formed by an S_N2 reaction of *n*-butyllithium with methylene chloride. Closs ruled out this possibility by showing that at $-30°$, where 1-pentene is formed rapidly from methylene chloride and *n*-butyllithium, *n*-pentyl chloride is essentially inert to *n*-butyllithium (215). Therefore the most plausible mechanism for the reaction is as shown below (the formation of *n*-butylmethylene may be a concerted reaction not involving the 1-chloropentyllithium intermediate shown).

$$CH_2Cl_2 + n\text{-}C_4H_9Li \rightarrow H{-}\ddot{C}{-}Cl + n\text{-}C_4H_{10} + LiCl$$

$$H{-}\ddot{C}{-}Cl + n\text{-}C_4H_9Li \rightarrow n\text{-}C_4H_9{-}\underset{\substack{|\\Li}}{CH}Cl \xrightarrow{-LiCl} CH_3CH_2CH_2CH_2{-}\ddot{C}{-}H$$

$$CH_3CH_2CH_2CH_2{-}\ddot{C}{-}H \rightarrow CH_3CH_2{-}CH{-}CH_2 \text{ (cyclopropane ring closed by } CH_2\text{)}$$

$$CH_3CH_2CH_2CH_2{-}\ddot{C}{-}H \rightarrow CH_3CH_2CH_2CH{=}CH_2$$

The formation of ethylcyclopropane, as shown in the reaction scheme, provides added evidence for the intermediacy of an alkylmethylene. Closs studied the reaction of other alkyllithium compounds with methylene chloride and in several cases obtained about the same product mixtures as those observed when the corresponding alkylmethylenes are generated from tosylhydrazones and base.

Kirmse described evidence that alkylchloromethylenes react with alkyllithium compounds to give dialkylmethylenes (128).

DECOMPOSITION OF AZINES

The decomposition of azines, like that of diazo compounds, may be written as beginning with the formation of the very stable molecule nitrogen and a methylene.

$$RCH{=}N{-}N{=}CHR \rightarrow 2R{-}C{-}H + N_2$$

In the case of azines, however, only half as many molecules of nitrogen are formed per molecule of methylene. This is probably the principal reason why the decomposition of azines has not yet been developed into an efficient method for the generation of methylenes.

Rice and Glasebrook reported that ethylene, nitrogen, hydrogen cyanide, a non-volatile oil, and probably methane were formed in the pyrolysis of acetaldazine. They suggested that the ethylene was formed by the isomerization of ethylidene but they could not prove the intermediacy of ethylidene by mirror-removal experiments. Brinton found that the photolysis of acetaldazine yields a little 2-butene and more ethylene, but under the conditions he used, acetonitrile and ammonia were the principal reaction products, showing that most of the reaction did not involve cleavage of the carbon-nitrogen double bonds (216).

The pyrolysis of benzalazines has long been known to yield stilbene derivatives, but, according to Zimmerman and Somasekhara, this reaction is probably an ionic chain process, in which methylenes play no part (in the chain-propagation steps at least) (217).

FORMATION OF ETHYLIDENE FROM ETHANE

Wijnen showed that the principal products of the photolysis of ethane (this must be carried out in the vacuum ultraviolet region) are hydrogen, methane, propane, butane, and ethylene.

Okabe and McNesby found that when a mixture of CH_3CH_3 and CD_3CD_3 are photolyzed the resultant hydrogen is almost entirely a mixture of H_2 and D_2, with no more HD being formed than could be accounted for by the presence of a little CD_3CD_2H (218). This shows that the hydrogen formed was detached as molecules from the reactants and rules out any mechanism such as the following, in which the hydrogen comes off one atom at a time.

$$CH_3CH_3 \xrightarrow{h\nu} CH_3CH_2\cdot + H\cdot$$
$$H\cdot + CH_3CH_3 \rightarrow H_2 + CH_3CH_2\cdot$$

When CH_3CD_3 was used as the reactant the hydrogen produced was found to be 58 per cent H_2, 17 per cent HD, and 25 per cent D_2, showing that about 83 per cent of the reaction involves the removal of two hydrogen atoms from the same carbon atom to give ethylidene and about 17 per cent involves the loss of two hydrogen atoms from adjacent carbon atoms to give ethylene.

$$CH_3CH_3 \xrightarrow{h\nu} CH_3\text{—}C\text{—}H + H_2$$
$$CH_3CH_3 \xrightarrow{h\nu} CH_2\text{=}CH_2 + H_2$$

Analysis of the deuteriated methanes produced in the reactions described shows that the ethane is also decomposing by another path, involving a transformation to a molecule of methane and one of methylene.

$$CH_3CH_3 \xrightarrow{h\nu} CH_4 + CH_2$$

The propane and butane observed are believed to arise from the attack of methylene and ethylidene on the ethane, which was always present in large excess since the reactions were carried only to a small fraction of completion.

Hydrogen has also been shown to be removed molecularly from ethane in the high-energy-electron-radiolysis of ethane, but in this case there may be electrically charged intermediates formed.

References

1. DOERING, W. v. E., and L. H. KNOX, "The Reaction of Carbalkoxycarbene with Saturated Hydrocarbons," *J. Am. Chem. Soc.*, **78**, 4947 (1956). The origin of the term "carbene" is described in Footnote 9 of this article.
2. KNUNYANTS, I. L., N. P. GAMBARYAN, and E. M. ROKHLIN, "Karbeny," *Uspekhi Khim.*, **27**, 1361–1436 (1958); a review with 545 references.
3. KIRMSE, W., "Reaktionen mit Carbenen und Iminen als Zwischenstufen," *Angew. Chem.*, **71**, 537 (1959); a review with 58 references.
4. KIRMSE, W., "Neues über Carbene," *Angew. Chem.*, **73**, 161 (1961); a review with 91 references.

4a. MIGINIAC, P., "Les carbènes," *Bull. soc. chim. France*, 2000–2014 (1962); a review with 357 references.

4b. CHINOPOROS, E., "Carbenes. Reactive Intermediates Containing Divalent Carbon," *Chem. Revs.*, **63**, 235–255 (1963); a review with 153 references.

5. HUISGEN, R., "Altes und Neues über aliphatische Diazoverbindungen," *Angew. Chem.*, **67**, 439 (1955); a review with 209 references.
6. ZOLLINGER, H., *Azo and Diazo Chemistry, Aliphatic and Aromatic Compounds*, Interscience Publishers, Inc., New York, 1961, especially Chapters 5 and 6.

6a. PARHAM, W. E., and E. E. SCHWEIZER, "Halocyclopropanes from Halocarbenes," *Organic Reactions*, **13**, 55–90 (1963); a review with 140 references.

7. HINE, J., "Methylenes," Chapter 24 in *Physical Organic Chemistry*, 2nd ed., McGraw-Hill Book Co., Inc., New York, 1962.
8. STAUDINGER, H., and O. KUPFER, "Uber Reaktionen des Methylens. III. Diazomethan," *Ber.*, **45**, 501 (1912).
9. RICE, F. O., and A. L. GLASEBROOK, "The Thermal Decomposition of Organic Compounds from the Standpoint of Free Radicals. XI. The Methylene Radical," *J. Am. Chem. Soc.*, **56**, 2381 (1934).
10. PEARSON, T. G., R. H. PURCELL, and G. S. SAIGH, "Methylene," *J. Chem. Soc.*, 409 (1938).
11. HERZBERG, G., "The Spectra and Structures of Free Methyl and Free Methylene," *Proc. Roy. Soc. (London)*, **A262**, 291 (1961).

12. Strachan, A. N., and W. A. Noyes, Jr., "Photochemical Studies. XLIX. Ketene and Ketene-Oxygen Mixtures," *J. Am. Chem. Soc.*, **76,** 3258 (1954).
13. Kistiakowsky, G. B., and N. W. Rosenberg, "Photochemical Decomposition of Ketene. II," *J. Am. Chem. Soc.*, **72,** 321 (1950).
14. Kassel, L. S., "The Thermal Decomposition of Methane," *J. Am. Chem. Soc.*, **54,** 3949 (1932); "The Role of Methyl and Methylene Radicals in the Decomposition of Methane," *J. Am. Chem. Soc.*, **57,** 833 (1935).
15. Belchetz, L., "The Thermal Catalytic Decomposition of Methane," *Trans. Faraday Soc.*, **30,** 170 (1934).
16. Rice, F. O., and M. D. Dooley, "The Thermal Decomposition of Organic Compounds from the Standpoint of Free Radicals. XII. The Decomposition of Methane," *J. Am. Chem. Soc.*, **56,** 2747 (1934).
17. Gevantman, L. H., and R. R. Williams, Jr., "Detection and Identification of Free Radicals in the Radiolysis of Alkanes and Alkyl Iodides," *J. Phys. Chem.*, **56,** 569 (1952).
18. Letort, M., and X. Duval, "Radicaux libres dans le méthane soumis à la décharge électrique," *Compt. rend.*, **219,** 452 (1944).
18a. Ausloos, P. J., and S. G. Lias, "Radiolysis of Methane," *J. Chem. Phys.*, **38,** 2207 (1963).
19. Mahan, B. H., and R. Mandal, "Vacuum Ultraviolet Photolysis of Methane," *J. Chem. Phys.*, **37,** 207 (1962).
20. Chanmugam, J., and M. Burton, "Reactions of Free Methylene: Photolysis of Ketene in Presence of Other Gases," *J. Am. Chem. Soc.*, **78,** 509 (1956).
21. Bell, J. A., and G. B. Kistiakowsky, "The Reactions of Methylene. VI. The Addition of Methylene to Hydrogen and Methane," *J. Am. Chem. Soc.*, **84,** 3417 (1962).
21a. Prophet, H., "Heat of Formation of Methylene," *J. Chem. Phys.*, **38,** 2345 (1963).
22. Meerwein, H., H. Rathjen, and H. Werner, "Die Methylierung von RH-Verbindungen mittels Diazomethan unter Mitwirkung des Lichtes," *Ber.*, **75,** 1610 (1942).
23. Frey, H. M., and G. B. Kistiakowsky, "Reactions of Methylene. I. Ethylene, Propane, Cyclopropane, and *n*-Butane," *J. Am. Chem. Soc.*, **79,** 6373 (1957).
24. Frey, H. M., "The Abstraction Reactions of Methylene," *Proc. Chem. Soc.*, 318 (1959).
25. Doering, W. v. E., and H. Prinzbach, "Mechanism of Reaction of Methylene with the Carbon-Hydrogen Bond. Evidence for Direct Insertion," *Tetrahedron*, **6,** 24 (1959).
26. Doering, W. v. E., R. G. Buttery, R. G. Laughlin, and N. Chaudhuri, "Indiscriminate Reaction of Methylene with the Carbon-Hydrogen Bond," *J. Am. Chem. Soc.*, **78,** 3224 (1956).
27. Richardson, D. B., M. C. Simmons, and I. Dvoretzky, "The Reactivity of Methylene from Photolysis of Diazomethane," *J. Am. Chem. Soc.*, **83,** 1934 (1961).

28. Frey, H. M., "The Addition of Methylene to Cyclobutane and the Decomposition of Excited Methylcyclobutane," *Trans. Faraday Soc.*, **56**, 1201 (1960).
29. Frey, H. M., and I. D. R. Stevens, "The Formation of Methylene by the Photolysis of Diazirine (Cyclodiazomethane)," *Proc. Chem. Soc.*, 79 (1962).
30. Skell, P. S., and R. C. Woodworth, "Structure of Carbene, CH_2," *J. Am. Chem. Soc.*, **78**, 4496 (1956).
31. Doering, W. v. E., and P. LaFlamme, "The *cis* Addition of Dibromocarbene and Methylene to *cis*- and *trans*-Butene," *J. Am. Chem. Soc.*, **78**, 5447 (1956).
32. Frey, H. M., "Reactions of Vibrationally Excited Molecules. I. The Reaction of Methylene with *iso*-Butene," *Proc. Roy. Soc. (London)*, **A250**, 409 (1959).
33. Frey, H. M., "Reactions of Vibrationally Excited Molecules. II. The Reaction of Methylene with *trans*- and *cis*-Butene-2," *Proc. Roy. Soc. (London)*, **A251**, 575 (1959).
34. Anet, F. A. L., R. F. W. Bader, and A.-M. Van der Auwera, "Chemical Evidence for a Triplet Ground State for Methylene," *J. Am. Chem. Soc.*, **82**, 3217 (1960).
35. Frey, H. M., "Chemical Evidence for the Ground State of Methylene," *J. Am. Chem. Soc.*, **82**, 5947 (1960).
36. Kopecky, K. R., G. S. Hammond, and P. A. Leermakers, "The Triplet State of Methylene in Solution," *J. Am. Chem. Soc.*, **83**, 2397 (1961); **84**, 1015 (1962).
37. Frey, H. M., "Addition of Methylene to Butadiene and the Unimolecular Isomerization of Excited Vinylcyclopropane," *Trans. Faraday Soc.*, **58**, 516 (1962).
38. Franzen, V., "Untersuchungen über Carbene, X Notiz über die 1.4-Addition von Methylen an Butadien," *Chem. Ber.*, **95**, 571 (1962).
39. Simmons, H. E., and R. D. Smith, "A New Synthesis of Cyclopropanes from Olefins," *J. Am. Chem. Soc.*, **80**, 5323 (1958); **81**, 4256 (1959).
40. Wittig, G., and K. Schwarzenbach, "Diazomethan und Zinkjodid," *Angew. Chem.*, **71**, 652 (1959).
41. Friedman, L., and J. G. Berger, "Carbene by the Dehydrohalogenation of Methyl Chloride," *J. Am. Chem. Soc.*, **82**, 5758 (1960).
42. Franzen, V., and G. Wittig, "Trimethylammonium-methylid als Methylen-Donator," *Angew. Chem.*, **72**, 417 (1960).
43. Wittig, G., and R. Polster, "Uber die Struktur der Stickstoff-ylide," *Ann.*, **599**, 1 (1956).
44. Goldfinger, P., P. Le Goff, and M. Letort, "Essais d'identification simultanée du radical libre méthylène par les methodes chimique et spectrographique," *J. chim. phys.*, **47**, 866 (1950).
45. Bawn, C. E. H., and C. F. H. Tipper, "The Reaction of Free Alkyl Radicals in the Gas Phase," *Disc. Faraday Soc.*, **2**, 104 (1947).
46. Wilson, T. B., and G. B. Kistiakowsky, "Reactions of Methylene.

III. Addition to Carbon Monoxide," *J. Am. Chem. Soc.*, **80,** 2934 (1958).

47. URRY, W. H., and J. R. EISZNER, "Photochemical Reactions of Diazomethane with Polyhalomethanes and α-Haloesters," *J. Am. Chem. Soc.*, **73,** 2977 (1951); **74,** 5822 (1952).
48. FRANZEN, V., "Reaktionen von Carbenen mit Alkylhalogeniden. Eine einfache Synthese von Neopentylchlorid und Neopentylbromid," *Ann.*, **627,** 22 (1959).
49. BRADLEY, J. N., and A. LEDWITH, "The Reaction of Carbene with Alkyl Halides," *J. Chem. Soc.*, 1495 (1961).
50. LEMMON, R. M., and W. STROHMEIER, "The Distribution of Radioactivity in Toluene Formed from Benzene and Photolyzed Diazomethane-C^{14}," *J. Am. Chem. Soc.*, **81,** 106 (1959).
51. DOERING, W. v. E., and L. H. KNOX, "Synthesis of Tropolone," *J. Am. Chem. Soc.*, **72,** 2305 (1950).
52. EVANS, M. V., and R. C. LORD, "Vibrational Spectra and Structure of the Tropilidene Molecule," *J. Am. Chem. Soc.*, **82,** 1876 (1960).
53. HINE, J., "Carbon Dichloride as an Intermediate in the Basic Hydrolysis of Chloroform. A Mechanism for Substitution Reactions at a Saturated Carbon Atom," *J. Am. Chem. Soc.*, **72,** 2438 (1950).
54. HINE, J., and A. M. DOWELL, JR., "Carbon Dihalides as Intermediates in the Basic Hydrolysis of Haloforms. III. Combination of Carbon Dichloride with Halide Ions," *J. Am. Chem. Soc.*, **76,** 2688 (1954).
55. HORIUTI, J., K. TANABE, K. TANAKA, and M. KATAYAMA, "The Mechanism of the Decomposition of Chloroform," *J. Research Inst. Catalysis, Hokkaido Univ.*, **3,** 119, 147 (1955); **6,** 44, 57 (1958); **7,** 79 (1959); **8,** 12 (1960).
56. HINE, J., and S. J. EHRENSON, "The Effect of Structure on the Relative Stability of Dihalomethylenes," *J. Am. Chem. Soc.*, **80,** 824 (1958).
57. HINE, J., and P. B. LANGFORD, "Methylene Derivatives as Intermediates in Polar Reactions. IX. The Concerted Mechanism for α-Eliminations of Haloforms," *J. Am. Chem. Soc.*, **79,** 5497 (1957).
58. DOERING, W. v. E., and A. K. HOFFMANN, "The Addition of Dichlorocarbene to Olefins," *J. Am. Chem. Soc.*, **76,** 6162 (1954).
59. SKELL, P. S., and A. Y. GARNER, "The Stereochemistry of Carbene-Olefin Reactions. Reactions of Dibromocarbene with the *cis*- and *trans*-2-Butenes," *J. Am. Chem. Soc.*, **78,** 3409 (1956).
60. SKELL, P. S., and A. Y. GARNER, "Reaction of Bivalent Carbon Compounds. Reactivities in Olefin-Dibromocarbene Reactions," *J. Am. Chem. Soc.*, **78,** 5430 (1956).
61. DOERING, W. v. E., and W. A. HENDERSON, JR., "The Electron-seeking Demands of Dichlorocarbene in its Addition to Olefins," *J. Am. Chem. Soc.*, **80,** 5274 (1958).
62. VENKATESWARLU, P., "On the Emission Bands of CF_2," *Phys. Rev.*, **77,** 676 (1950).

63. Laird, R. K., E. B. Andrews, and R. F. Barrow, "The Absorption Spectrum of CF_2," *Trans. Faraday Soc.*, **46**, 803 (1950).
64. Mann, D. E., and B. A. Thrush, "On the Absorption Spectrum of CF_2 and its Vibrational Analysis," *J. Chem. Phys.*, **33**, 1732 (1960).
65. Brewer, L., J. L. Margrave, R. F. Porter, and K. Wieland, "The Heat of Formation of CF_2," *J. Phys. Chem.*, **65**, 1913 (1961).
66. Wagner, W. M., "A New Synthesis of Dichlorocarbene," *Proc. Chem. Soc.*, 229 (1959).
67. Wagner, W. M., H. Kloosterziel, and S. van der Ven, "The Thermal Decarboxylation of Alkali Trichloroacetates in Aprotic Solvents," *Rec. trav. chim.*, **80**, 740 (1961).
68. Hine, J., and D. C. Duffey, "Methylene Derivatives as Intermediates in Polar Reactions. XVI. The Decomposition of Chlorodifluoroacetic Acid," *J. Am. Chem. Soc.*, **81**, 1131 (1959).
69. Kaneko, Y., and Y. Sato, "Homogeneous Hydrogen Exchange between Chloroform and Heavy Water," *J. Research Inst. for Catalysis, Hokkaido Univ.*, **6**, 28 (1958).
70. Birchall, J. M., G. W. Cross, and R. N. Haszeldine, "Difluorocarbene," *Proc. Chem. Soc.*, 81 (1960).
71. Parham, W. E., and F. C. Loew, "Formation of Carbenes from α-Haloesters," *J. Org. Chem.*, **23**, 1705 (1958).
72. Parham, W. E., F. C. Loew, and E. E. Schweizer, "Mechanism of Carbene Formation from *t*-Butyl Dichloroacetate," *J. Org. Chem.*, **24**, 1900 (1959).
73. Kadaba, P. K., and J. O. Edwards, "Hexachloroacetone as a Novel Source of Dichlorocarbene," *J. Org. Chem.*, **25**, 1431 (1960).
74. Grant, F. W., and W. B. Cassie, "Hexachloroacetone as a Source of Dichlorocarbene," *J. Org. Chem.*, **25**, 1433 (1960).
75. Miller, W. T., Jr., and C. S. Y. Kim, "Reactions of Alkyllithiums with Polyhalides," *J. Am. Chem. Soc.*, **81**, 5908 (1959).
76. Franzen, V., "Untersuchungen über Carbene, XII Bestimmung der Lebensdauer des Difluorcarbens," *Chem. Ber.*, **95**, 1964 (1962).
77. Hauser, C. R., W. G. Kofron, W. R. Dunnavant, and W. F. Owens, "Reactions of Alkali Diphenylmethides with Certain Polyhalides. Displacement on Halogen or Hydrogen," *J. Org. Chem.*, **26**, 2627 (1961).
78. Semeluk, G. P., and R. B. Bernstein, "The Thermal Decomposition of Chloroform. I. Products," *J. Am. Chem. Soc.*, **76**, 3793 (1954); "II. Kinetics," *J. Am. Chem. Soc.*, **79**, 46 (1957).
79. Shilov, A. E., and R. D. Sabirova, "Mekhanizm i Izotopnyi Effekt Pervichnovo Akta Termicheskovo Raspada Khloroforma," *Doklady Akad. Nauk*, **114**, 1058 (1957); "The Mechanism of the First Stage in the Thermal Decomposition of Chloromethanes. II. The Decomposition of Chloroform," *Russ. J. Phys. Chem.*, **34**, 408 (1960).
80. Shilov, A. E., and R. D. Sabirova, "Mekhanizm Pervichnovo Akta Termicheskovo Raspada Khlorproizvodnykh Metana. I.

Raspad Chetyrekhkhloristovo Ugleroda i Khloristovo Metila," *Zhur. Fiz. Khim.*, **33**, 1365 (1959).

81. HINE, J., and J. J. PORTER, "The Formation of Difluoromethylene from Difluoromethyl Phenyl Sulfone and Sodium Methoxide," *J. Am. Chem. Soc.*, **82**, 6178 (1960).
82. SEYFERTH, D., R. J. MINASZ, A. J.-H. TREIBER, J. M. BURLITCH, and S. R. DOWD, "The Reaction of Phenyl(trihalomethyl)mercurials with Olefins of Low Reactivity toward Dihalocarbenes," *J. Org. Chem.*, **28**, 1163 (1963).
83. BADEA, F., and C. D. NENITZESCU, "Neue Methode zur Darstellung von Carbenen," *Angew. Chem.*, **72**, 415 (1960); V. IOAN, F. BADEA, E. CIORANESCU, and C. D. NENITZESCU, "Dichlorcarben beim thermischen Zerfall von Silbertrichloracetat," *Angew. Chem.*, **72**, 416 (1960).
84. DYAKONOV, I. A., I. A. FAVORSKAYA, L. P. DANILKINA, and E. M. AUVINEN, "The Reaction of Dichlorocarbene with Enyne Hydrocarbons," *J. Gen. Chem. U. S. S. R. (English translation)*, **30**, 3475 (1960).
85. KURSANOV, D. N., M. E. VOLPIN, and YU. D. KORESHKOV, "Diphenylcyclopropenone, a Three-membered Analog of Tropone," *Bull. Acad. Sci. U. S. S. R., Div. Chem. S. S. R. (English translation)*, 535 (1959); "Reaction of Dihalocarbenes with Tolan. Synthesis of Diphenylcyclopropenone and Diphenylhydroxycyclopropenylium Salts," *J. Gen. Chem. U. S. S. R. (English translation)*, **30**, 2855 (1960).
86. BRESLOW, R., and R. PETERSON, "Dipropylcyclopropenone," *J. Am. Chem. Soc.*, **82**, 4426 (1960).
87. MCELVAIN, S. M., and P. L. WEYNA, "Ketene Acetals. XXXVII. Cyclopropanone Acetals from Ketone Acetals and Carbenes," *J. Am. Chem. Soc.*, **81**, 2579 (1959).
88. BALL, W. J., and S. R. LANDOR, "The Addition of Carbenes to Allenes," *Proc. Chem. Soc.*, 246 (1961).
89. FIELDS, E. K., and J. M. SANDRI, "Addition of Dihalocarbenes to Imines," *Chem. & Ind. (London)*, 1216 (1959).
90. CIAMICIAN, G. L., and M. DENNSTEDT, "Ueber die Einwirkung des Chloroforms auf die Kaliumverbindung Pyrrols," *Ber.*, **14**, 1153 (1881); "Studien über Verbindungen aus der Pyrrolreihe. Ueberführung des Pyrrols in Pyridin," *Ber.*, **15**, 1172 (1882).
91. PARHAM, W. E., and H. E. REIFF, "Ring Expansion during the Reaction of Indenylsodium and Chloroform," *J. Am. Chem. Soc.*, **77**, 1177 (1955).
92. PARHAM, W. E., H. E. REIFF, and P. SWARZENTRUBER, "The Formation of Naphthalenes from Indenes. II," *J. Am. Chem. Soc.*, **78**, 1437 (1956).
93. MURRAY, R. W., "The Reaction of Dichlorocarbene with Anthracene," *Tetrahedron Letters*, no. 7, 27 (1960).
94. PARHAM, W. E., D. A. BOLON, and E. E. SCHWEIZER, "The Reaction of Halocarbenes with Aromatic Systems: Synthesis of Chlorotropones," *J. Am. Chem. Soc.*, **83**, 603 (1961).

95. SKELL, P. S., and S. R. SANDLER, "Reactions of 1,1-Dihalocyclopropanes with Electrophilic Reagents. Synthetic Route for Inserting a Carbon Atom between the Atoms of a Double Bond," *J. Am. Chem. Soc.*, **80**, 2024 (1958).
96. VOGEL, E., "Addition von Carbenen an Cyclooctatetraen," *Angew. Chem.*, **73**, 548 (1961).
97. NEUREITER, N. P., "Pyrolysis of 1,1-Dichloro-2-vinylcyclopropane. Synthesis of 2-Chlorocyclopentadiene," *J. Org. Chem.*, **24**, 2044 (1959).
98. WYNBERG, H. E., "Synthesis of Cycloheptatriene," *J. Org. Chem.*, **24**, 264 (1959).
99. WYNBERG, H., "The Reimer-Tiemann Reaction," *Chem. Revs.*, **60**, 169 (1960); review with 144 references.
100. HINE, J., and J. M. VAN DER VEEN, "The Mechanism of the Reimer-Tiemann Reaction," *J. Am. Chem. Soc.*, **81**, 6446 (1959).
101. HINE, J., and J. M. VAN DER VEEN, "Methylene Derivatives as Intermediates in Polar Reactions. XXIV. The Reimer-Tiemann Reaction with Two *p*-Substituted Phenols," *J. Org. Chem.*, **26**, 1406 (1961).
102. SMITH, P. A. S., and N. W. KALENDA, "Investigation of Some Dialkylamino Isocyanides," *J. Org. Chem.*, **23**, 1599 (1958).
103. SAUNDERS, M., and R. W. MURRAY, "The Reaction of Dichlorocarbene with Amines," *Tetrahedron*, **6**, 88 (1959); **11**, 1 (1960).
104. CLEMENS, D. H., E. Y. SHROPSHIRE, and W. D. EMMONS, "Orthoamides and Formamidinium Salts," *J. Org. Chem.*, **27**, 3664 (1962).
105. SPEZIALE, A. J., G. J. MARCO, and K. W. RATTS, "A Novel Synthesis of 1,1-Dihaloolefins," *J. Am. Chem. Soc.*, **82**, 1260 (1960).
106. SEYFERTH, D., S. O. GRIM, and T. O. READ, "A New Preparation of Triphenylphosphinemethylenes by the Reaction of Carbenes with Triphenylphosphine," *J. Am. Chem. Soc.*, **82**, 1510 (1960).
107. PARHAM, W. E., and R. KONCOS, "The Reaction of Dichlorocarbene with 2H-1-Benzothiopyran and 4H-1-Benzothiopyran," *J. Am. Chem. Soc.*, **83**, 4034 (1961).
108. FIELDS, E. K., "Insertion of Dichlorocarbene into Aromatic Hydrocarbons," *J. Am. Chem. Soc.*, **84**, 1744 (1962).
108a. SEYFERTH, D., and J. M. BURLITCH, "The Preparation of Dihalomethyl Derivatives of Carbon, Silicon and Germanium by the Action of Phenyl(trihalomethyl)mercurials on C—H, Si—H, and Ge—H Linkages," *J. Am. Chem. Soc.*, **85**, 2667 (1963).
109. SCHÖLLKOPF, U., A. LERCH, and W. PITTEROFF, "Phenoxy-, Methoxy-, Butoxy-, und Isopropoxy-Carben," *Tetrahedron Letters*, 241 (1962); and references cited therein.
110. HINE, J., and K. TANABE, "Isopropoxyfluoromethylene," *J. Am. Chem. Soc.*, **79**, 2654 (1957); **80**, 3002 (1958).
111. HINE, J., A. D. KETLEY, and K. TANABE, "Methylene Derivatives as Intermediates in Polar Reactions. XIX. The Reaction of Potassium Isopropoxide with Chloroform, Bromoform, and Dichlorofluoromethane," *J. Am. Chem. Soc.*, **82**, 1398 (1960).

112. Cleaver, C. S., "1,2-Difluoro-1,2-dialkoxyethylenes and Preparation Thereof," U. S. Patent 2,853,531.
113. Hine, J., E. L. Pollitzer, and H. Wagner, "The Dehydration of Alcohols in the Presence of Haloforms and Alkali," *J. Am. Chem. Soc.*, **75,** 5607 (1953).
114. Skell, P. S., and I. Starer, "Reaction of Alkoxide with CX_2 to Produce Carbonium Ion Intermediates," *J. Am. Chem. Soc.*, **81,** 4117 (1959).
115. Hine, J., R. J. Rosscup, and D. C. Duffey, "Formation of the Intermediate Methoxychloromethylene in the Reaction of Dichloromethyl Methyl Ether with Base," *J. Am. Chem. Soc.*, **82,** 6120 (1960).
116. McBee, E. T., J. D. Idol, Jr., and C. W. Roberts, "Chemistry of Hexachlorocyclopentadiene. VI. Diels-Alder Adduct with Alkynes," *J. Am. Chem. Soc.*, **77,** 6674 (1955).
116a. Corey, E. J., and R. A. E. Winter, "A New, Stereospecific Olefin Synthesis from 1,2-Diols," *J. Am. Chem. Soc.*, **85,** 2677 (1963).
117. Schöllkopf, U., and G. J. Lehmann, "Phenylmercaptocyclopropane aus Phenylmercapto-carben und Olefinen," *Tetrahedron Letters*, 165 (1962).
118. Hine, J., and J. J. Porter, "Methylenes as Intermediates in Polar Reactions. XXI. A Sulfur-containing Methylene," *J. Am. Chem. Soc.*, **82,** 6118 (1960).
119. Arens, J. F., "Some Aspects of the Chemistry of Organic Sulfides," Chapter 23 in *Organic Sulfur Compounds*, N. Kharasch, Editor, Pergamon Press, New York, 1961.
120. Hine, J., R. P. Bayer, and G. G. Hammer, "Formation of Bis-(methylthio)methylene from Methyl Orthothioformate and Potassium Amide," *J. Am. Chem. Soc.*, **84,** 1751 (1962).
121. Volpin, M. E., V. G. Dulova, and D. N. Kursanov, "Obrazovanie Tropiliya pri Deistvii Monogalokarbenov na Benzol," *Doklady Akad. Nauk S. S. S. R.*, **128,** 951 (1959); "New Mutual Conversion of Aromatic Systems. Tropilium Salts and Benzene," *Tetrahedron*, **8,** 33 (1960).
122. Closs, G. L., and L. E. Closs, "Synthesis of Chlorocyclopropanes from Methylene Chloride and Olefins," *J. Am. Chem. Soc.*, **81,** 4996 (1959).
123. Closs, G. L., and L. E. Closs, "Carbenes from Alkyl Halides and Organolithium Compounds. I. Synthesis of Chlorocyclopropanes," *J. Am. Chem. Soc.*, **82,** 5723 (1960).
123a. Closs, G. L., R. A. Moss, and J. J. Coyle, "Steric Course of Some Carbenoid Additions to Olefins," *J. Am. Chem. Soc.*, **84,** 4985 (1962).
124. Closs, G. L., and L. E. Closs, "Addition of Chlorocarbene to Benzene," *Tetrahedron Letters*, no. 10, 38 (1960).
125. Closs, G. L., and L. E. Closs, "Carbenes from Alkyl Halides and Organolithium Compounds. III. Syntheses of Alkyltropones from Phenols," *J. Am. Chem. Soc.*, **83,** 599 (1961).

126. CLOSS, G. L., and G. M. SCHWARTZ, "Carbenes from Alkyl Halides and Organolithium Compounds. II. The Reactivity of Chlorocarbene in its Addition to Olefins," *J. Am. Chem. Soc.*, **82,** 5729 (1960).
126a. CLOSS, G. L., and J. J. COYLE, "The Preparation, Pyrolysis and Photolysis of Chlorodiazomethane," *J. Am. Chem. Soc.*, **84,** 4350 (1962).
127. BRESLOW, R., R. HAYNIE, and J. MIRRA, "The Synthesis of Diphenylcyclopropenone," *J. Am. Chem. Soc.*, **81,** 247 (1959).
128. KIRMSE, W., "Carbene in Reaktionen metallorganischer Verbindungen," *Angew. Chem.*, **74,** 183 (1962), and earlier references cited therein.
129. HASZELDINE, R. N., and J. C. YOUNG, "α-Elimination and Carbene Formation from Silicon Compounds," *Proc. Chem. Soc.*, 394 (1959).
130. BRUYLANTS, P., "Carbylamines," *Grignard's Traité de Chimie Organique*, Vol. XIII, 1941, pp. 847–868.
131. BOTHNER-BY, A. A., "Stereochemistry of the Rearrangement Involving Migration between Multiply-bonded Carbons," *J. Am. Chem. Soc.*, **77,** 3293 (1955).
132. CURTIN, D. Y., E. W. FLYNN, and R. F. NYSTROM, "Reaction of Stereoisomeric C^{14}-Labeled 1-Bromo-2,2-diarylethylenes and β-Bromostyrenes with Butyllithium," *J. Am. Chem. Soc.*, **80,** 4599 (1958).
133. PRITCHARD, J. G., and A. A. BOTHNER-BY, "Base-Initiated Dehydrohalogenation and Rearrangement of 1-Halo-2,2-diphenylethylenes in *t*-Butyl Alcohol. The Effect of Deuterated Solvent," *J. Phys. Chem.*, **64,** 1271 (1960).
134. HAUSER, C. R., and D. LEDNICER, "Reaction of 9-Bromomethylenefluorene with Potassium Amide in Liquid Ammonia. Dimerization," *J. Org. Chem.*, **22,** 1248 (1957).
135. CURTIN, D. Y., and W. H. RICHARDSON, "The Reactions of Exocyclic Vinyl Halides with Phenyllithium," *J. Am. Chem. Soc.*, **81,** 4719 (1959).
136. HENNION, G. F., and D. E. MALONEY, "The Hydrolysis of 3-Chloro-3-methyl-1-butyne and 1-Chloro-3-methyl-1,2-butadiene," *J. Am. Chem. Soc.*, **73,** 4735 (1951).
137. HENNION, G. F., and K. W. NELSON, "The Kinetics of the Hydrolysis of Acetylenic Chlorides and Their Reactions with Primary and Secondary Aliphatic Amines," *J. Am. Chem. Soc.*, **79,** 2142 (1957).
138. HENNION, G. F., and E. G. TEACH, "The Preparation of Some Acetylenic Primary Amines," *J. Am. Chem. Soc.*, **75,** 1653 (1953).
139. HARTZLER, H. D., "The Stereochemistry and Relative Rates of Addition of Dimethylvinylidene Carbene to Olefins," *J. Am. Chem. Soc.*, **83,** 4997 (1961), and earlier references cited therein.
140. SHINER, V. J., JR., and J. W. WILSON, "The Mechanisms of Substitution of Propargylic Halides. 3-Bromo-3-methyl-1-butyne," *J. Am. Chem. Soc.*, **84,** 2402 (1962).

141. NEWMAN, M. S., *et al.*, "New Reactions Involving Alkaline Treatment of 3-Nitroso-2-oxazolidones," *J. Am. Chem. Soc.*, **73,** 4199 (1951); **76,** 1840 (1954).
142. BAYES, K., "The Photolysis of Carbon Suboxide," *J. Am. Chem. Soc.*, **83,** 3712 (1961); **84,** 4077 (1962).
143. MACKAY, C., P. POLAK, H. E. ROSENBERG, and R. WOLFGANG, "The Reactions of Atomic Carbon with Ethylene," *J. Am. Chem. Soc.*, **84,** 308 (1962).
143a. SKELL, P. S., and L. D. WESCOTT, "Chemical Properties of C_3, a Dicarbene," *J. Am. Chem. Soc.*, **85,** 1023 (1963).
144. HANNA, S. B., Y. ISKANDER, and Y. RIAD, "The Influence of the Nitro-group upon Side-chain Reactivity. Part I. The Reaction between 4-Nitrobenzyl Chloride and Alkali," *J. Chem. Soc.*, 217 (1961).
145. SWAIN, C. G., and E. R. THORNTON, "Mechanism of α-Elimination by Hydroxide Ion on *p*-Nitrobenzylsulfonium Ion in Aqueous Solution," *J. Am. Chem. Soc.*, **83,** 4033 (1961).
146. KHARASCH, M. S., W. NUDENBERG, and E. K. FIELDS, "Synthesis of Polyenes. IV," *J. Am. Chem. Soc.*, **66,** 1276 (1944), and earlier references cited therein.
147. HAUSER, C. R., W. R. BRASEN, P. S. SKELL, S. W. KANTOR, and A. E. BRODHAG, "Dimeric Olefins *versus* Amines from α-Aryl Alkyl Halides with Alkali Amides in Liquid Ammonia. Intermediate Dimeric Halides and Their Dehydrohalogenation. Imines from Amines," *J. Am. Chem. Soc.*, **78,** 1653 (1956).
148. BRASEN, W. R., S. W. KANTOR, P. S. SKELL, and C. R. HAUSER, "Self-alkylation of α-Phenylethyl Chloride to Form Isomeric Dimeric Halides and Dimeric Olefins by Amide Ion. Isomerization of *cis*-α,α'-Dimethylstilbene to *trans* Isomer," *J. Am. Chem. Soc.*, **79,** 397 (1957).
149. CLOSS, G. L., and L. E. CLOSS, "Phenylcarbene from Benzyl Chloride and *n*-Butyllithium," *Tetrahedron Letters*, no. 24, 26 (1960).
150. SCHÖLLKOPF, U., and M. EISERT, "α-Eliminierung bei alkalimetallorganischen Verbindungen, I Phenylcarben aus Lithium-benzylphenyl-äther Ein Beitrag zum Mechanismus der Ätherspaltung durch starke Basen," *Ann.*, **664,** 76 (1963).
150a. BETHELL, D., "Kinetics and Mechanism of the Formation of Bifluorenylidene from 9-Bromofluorene in *t*-Butyl Alcohol," *J. Chem. Soc.*, 666 (1963).
151. FRANZEN, V., "Untersuchungen über Carbene, VII. Aminaustausch beim Trimethylammonium-9-fluorenylid," *Chem. Ber.*, **93,** 557 (1960).
152. LUCK, S. M., D. G. HILL, A. T. STEWART, JR., and C. R. HAUSER, "The Elimination Reaction of Alpha and Beta Deuterated *n*-Octyl Bromides with Potassium Amide in Liquid Ammonia," *J. Am. Chem. Soc.*, **81,** 2784 (1959).
153. KIRMSE, W., and W. v. E. DOERING, "Cyclopropanes from 1°-Alkyl Chlorides by α-Elimination," *Tetrahedron*, **11,** 266 (1960).

154. Friedman, L., and J. G. Berger, "Dehydrohalogenation of Neoalkyl Halides by Strong Base: Evidence of Carbene Intermediates," *J. Am. Chem. Soc.*, **83**, 500 (1961).
155. Cope, A. C., G. A. Berchtold, P. E. Peterson, and S. H. Sharman, "Proximity Effects. XXII. Evidence for the Mechanism of the Reaction of Medium-sized Ring Epoxides with Lithium Diethylamide," *J. Am. Chem. Soc.*, **82**, 6370 (1960).
156. Powell, J. W., and M. C. Whiting, "The Decomposition of Sulfonylhydrazone Salts. I. Mechanism and Stereochemistry," *Tetrahedron*, **7**, 305 (1959).
157. Friedman, L., and H. Shechter, "Carbenoid and Cationoid Decomposition of Diazo Hydrocarbons Derived from Tosylhydrazones," *J. Am. Chem. Soc.*, **81**, 5512 (1959); "Rearrangement and Fragmentation Reactions in Carbenoid Decomposition of Diazo Hydrocarbons," *J. Am. Chem. Soc.*, **82**, 1002 (1960); "Transannular and Hydrogen-Rearrangement Reactions in Carbenoid Decomposition of Diazocycloalkanes," *J. Am. Chem. Soc.*, **83**, 3159 (1961).
158. Weygand, F., and H. J. Bestmann, "Neuere präparative Methoden der organischen Chemie III Synthesen unter Verwendung von Diazoketonen," *Angew. Chem.*, **72**, 535 (1960); a review with 185 references.
159. Frey, H. M., and I. D. R. Stevens, "Hot Radical Effects in an Intramolecular Insertion Reaction," *J. Am. Chem. Soc.*, **84**, 2647 (1962).
160. Doering, W. v. E., and L. H. Knox, "Comparative Reactivity of Methylene, Carbomethoxycarbene and Bis-carboethoxycarbene toward the Saturated Carbon-Hydrogen Bond," *J. Am. Chem. Soc.*, **83**, 1989 (1961).
161. Gutsche, C. D., G. L. Bachman, and R. S. Coffey, "Chemistry of Bivalent Carbon Intermediates. IV. Comparative Intermolecular and Intramolecular Reactivities of Phenylcarbene to Various Bond Types," *Tetrahedron*, **18**, 617 (1962).
162. Parham, W. E., and W. R. Hasek, "Reactions of Diazo Compounds with Nitroölefins. IV. The Decomposition of Diphenyldiazomethane," *J. Am. Chem. Soc.*, **76**, 935 (1954).
163. Kirmse, W., L. Horner, and H. Hoffmann, "Über Lichtreaktionen IX Umsetzungen photochemisch erzeugter Carbene," *Ann.*, **614**, 19 (1958).
163a. Trozzolo, A. M., R. W. Murray, and E. Wasserman, "The Electron Paramagnetic Resonance of Phenylmethylene and Biphenylenemethylene; A Luminescent Reaction Associated with a Ground State Triplet Molecule," *J. Am. Chem. Soc.*, **84**, 4990 (1962).
164. Yates, P., and S. Danishefsky, "A Novel Type of Alkyl Shift," *J. Am. Chem. Soc.*, **84**, 879 (1962).
165. Lansbury, P. T., and J. G. Colson, "Intramolecular Carbene Insertion into a Carbon-Carbon Bond," *Chem. & Ind. (London)*, 821 (1962).

166. Doering, W. v. E., G. Laber, R. Vonderwahl, N. F. Chamberlain, and R. B. Williams, "The Structure of the Buchner Acids," *J. Am. Chem. Soc.*, **78**, 5448 (1956).
167. Alder, K., R. Muders, W. Krane, and P. Wirtz, "Über die Konstitution photochemisch dargestellter Norcaradien-carbonsäureester," *Ann.*, **627**, 59 (1959).
168. Treibs, W., *et al.*, "Über bi- und polycyclische Azulene," *Ann.*, **598**, 32, 38, 41 (1956).
169. Gutsche, C. D., and H. E. Johnson, "Experiments in the Colchicine Field. III. A New Method for the Synthesis of Tricyclic Fused Ring Structures," *J. Am. Chem. Soc.*, **77**, 5933 (1955).
170. Gutsche, C. D., E. F. Jason, R. S. Coffey, and H. E. Johnson, "Experiments in the Colchicine Field. V. The Thermal and Photochemical Decomposition of Various 2-(β-Phenylethyl)-phenyldiazomethanes and 2-(γ-Phenylpropyl)-phenyldiazomethanes," *J. Am. Chem. Soc.*, **80**, 5756 (1958).
171. Weygand, F., H. Dworschak, K. Koch, and S. Konstas, "Reaktionen des Trifluoracetyl-carbäthoxy-carbens. II. Mitteilung," *Angew. Chem.*, **73**, 409 (1961).
172. Doering, W. v. E., and T. Mole, "Cyclopropenes from Carbenes and Acetylenes. Stereo-selectivity in the Reaction of Carbomethoxycarbene with *cis*-Butene," *Tetrahedron*, **10**, 65 (1960).
173. Dyakonov, I. A., F. Gui-siya, G. L. Korichev, and M. I. Komendantov, "The Stereoselective Course of the Reaction between Carbethoxycarbene and the Stereoisomeric 1,2-Diphenylethenes," *J. Gen. Chem. U. S. S. R. (English Translation)*, **31**, 624 (1961).
174. Skell, P. S., and R. M. Etter, "Steric Discrimination in Reactions of Ethoxycarbonylcarbene: Norcarane-7-carboxylic Acid," *Proc. Chem. Soc.*, 443 (1961).
175. Etter, R. M., H. S. Skovronek, and P. S. Skell, "Diphenylmethylene, $(C_6H_5)_2C$, a Diradical Species," *J. Am. Chem. Soc.*, **81**, 1008 (1959).
175a. Doering, W. v. E., and M. Jones, Jr., "Light-Induced Interconversion of *cis* and *trans* 2,3-Dimethylspiro[cyclopropane-1,9[1]-Fluorene]," *Tetrahedron Letters*, 791 (1963).
176. Closs, G. L., and L. E. Closs, "Stereospecific Formation of Cyclopropanes by Reaction of Diphenyldibromomethane with Methyllithium and Olefins," *Angew. Chem. Intern. Ed. Engl.*, **1**, 334 (1962).
176a. Murray, R. W., A. M. Trozzolo, E. Wasserman, and W. A. Yager, "E.P.R. of Diphenylmethylene, a Ground-State Triplet," *J. Am. Chem. Soc.*, **84**, 3213 (1962).
176b. Brandon, R. W., G. L. Closs, and C. A. Hutchison, Jr., "Paramagnetic Resonance in Oriented Ground State Triplet Molecules," *J. Chem. Phys.*, **37**, 1878 (1963).
177. Closs, G. L., and L. E. Closs, "Carbenes from Alkyl Halides and Organolithium Compounds. V. Formation of Alkylcyclopro-

penes by Ring Closure of Alkenyl Substituted Carbenoid Intermediates," *J. Am. Chem. Soc.*, **85**, 99 (1963).
178. STORK, G., and J. FICINI, "Intramolecular Cyclization of Unsaturated Diazoketones," *J. Am. Chem. Soc.*, **83**, 4678 (1961).
179. HUISGEN, R., H. KÖNIG, G. BINSCH, and H. J. STURM, "1,3-Dipolare Additionen der Ketocarbene," *Angew. Chem.*, **73**, 368 (1961).
180. NOVÁK, J., J. RATUSKÝ, V. ŠNEBERK, and F. ŠORM, "Reaktionen von Diazoketonen. I. Die Reaktion des Diazoacetons mit ungesättigten Verbindungen," *Collection Czech. Chem. Commun.*, **22**, 1836 (1957).
181. JONES, W. M., "The Cyclopropyl Carbene," *J. Am. Chem. Soc.*, **82**, 6200 (1960).
181a. JONES, W. M., M. H. GRASLEY, and W. S. BREY, JR., "The Cyclopropylidene: Generation and Reactions," *J. Am. Chem. Soc.*, **85**, 2754 (1963).
182. SKELL, P. S., and J. KLEBE, "Structure and Properties of Propargylene, C_3H_2," *J. Am. Chem. Soc.*, **82**, 247 (1960).
183. DYAKONOV, I. A., and M. I. KOMENDANTOV, "Novaya Reaktsiya Diazouksusnovo Efira s Atsetilenovym Uglevodorodom Sintez Proizvodnovo Tsiklopropena," *Vestn. Leningr. Univ.*, **11**, No. 22, *Ser. Fiz. i Khim.* No. 4, 166 (1956); *Chem. Abstr.*, **52**, 2762i (1958); "Interaction of Diazoacetic Ester with Acetylenic Hydrocarbons," *J. Gen. Chem. U. S. S. R. (English translation)*, **29**, 1726 (1959).
184. BRESLOW, R., "Synthesis of the *s*-Triphenylcyclopropenyl Cation," *J. Am. Chem. Soc.*, **79**, 5318 (1957); R. BRESLOW and C. YUAN, "The *sym.*-Triphenylcyclopropenyl Cation, a Novel Aromatic System," *J. Am. Chem. Soc.*, **80**, 5991 (1958).
185. GUTSCHE, C. D., and M. HILLMAN, "Reactions of Ethyl Diazoacetate with Aromatic Compounds Containing Hetero Atoms Attached to the Benzyl Carbon," *J. Am. Chem. Soc.*, **76**, 2236 (1954).
186. DE GRAAFF, G. B. R., and G. VAN DE KOLK, "The Reaction between Ethyl Diazoacetate and 1-Butoxybutane," *Rec. trav. chim.*, **77**, 224 (1958).
187. KHARASCH, M. S., T. RUDY, W. NUDENBERG, and G. BÜCHI, "Reactions of Diazoacetates and Diazoketones. I. Reaction of Ethyl Diazoacetate with Cyclohexanone and with Acetone," *J. Org. Chem.*, **18**, 1030 (1953).
188. FRANZEN, V., and H. KUNTZE, "Untersuchungen über Carbene. IV. Reaktionen von Carbenen mit aliphatischen Aminen," *Ann.*, **627**, 15 (1959).
189. PHILLIPS, D. D., "The Reaction between Diazoacetic Ester and Allylic Halides," *J. Am. Chem. Soc.*, **76**, 5385 (1954).
190. URRY, W. H., and J. W. WILT, "Photochemical Reactions of Methyl Diazoacetate with Polyhalomethanes," *J. Am. Chem. Soc.*, **76**, 2594 (1954).
191. A detailed review of the Arndt-Eistert synthesis including its second step, the Wolff rearrangement, with emphasis on synthetic applications is given by W. E. BACHMANN and W. S. STRUVE,

"The Arndt-Eistert Synthesis," *Organic Reactions*, **1**, 38–62 (1942), and also References 5 and 158.

192. Horner, L., E. Spietschka, and A. Gross, "Zur Kenntnis der Umlagerungsvorgänge bei Diazo-ketonen, *o*-Chinondiaziden und Säureaziden," *Ann.*, **573**, 17 (1951); L. Horner and E. Spietschka, "Die präparative Bedeutung der Zersetzung von Diazocarbonylverbindungen im UV-Licht," *Chem. Ber.*, **85**, 225 (1952).
193. Franzen, V., "Untersuchungen über Carbene. II. Zur Existenz von Acetylenoxyden," *Ann.*, **614**, 31 (1958).
194. Horner, L., and E. Spietschka, "Über Lichtreaktionen. IV. Bicyclo-[1.1.2]-hexan-Derivate als Ergebnis der Umlagerung des Diazocamphers im Licht," *Chem. Ber.*, **88**, 934 (1955).
195. Süs, O., "Über die Natur der Belichtungsprodukte von diazoverbindungen. Übergänge von aromatischen 6-Ringen in 5-Ringe," *Ann.*, **556**, 65, 85 (1944); "Photosynthese des 5-Azaindols und 7-Azaindols," *Ann.*, **612**, 153 (1958).
196. Newman, M. S., and A. Arkell, "New Reactions on Decomposition of a Hindered α-Diazoketone," *J. Org. Chem.*, **24**, 385 (1959).
197. Franzen, V., "Eine Neue Methode zur Darstellung α,β-ungesättigter Ketone. Zerfall der Diazaketone R—CO—CN_2—CH_2—R′," *Ann.*, **602**, 199 (1957).
198. Meinwald, J., and P. G. Gassman, "Highly Strained Bicyclic Systems. I. The Synthesis of Some Bicyclo[2.1.1]hexanes of Known Stereochemistry," *J. Am. Chem. Soc.*, **82**, 2857 (1960).
199. Weygand, F., and K. Koch, "Eine neue Umlagerungsreaktion. Dichlormaleinsäure-äthylester-chloride aus Trichloracetyldiazoessigester," *Angew. Chem.*, **73**, 531 (1961).
200. Fields, R., and R. N. Haszeldine, "Rearrangement of Fluoroalkyl Carbenes," *Proc. Chem. Soc. (London)*, 22 (1960).
201. Kirmse, W., and L. Horner, "Über Lichtreaktionen. VIII. Photolyse von 1,2,3-Thiodiazolen," *Ann.*, **614**, 4 (1958).
202. DeMore, W. B., H. O. Pritchard, and N. Davidson, "Photochemical Experiments in Rigid Media at Low Temperatures. II. The Reactions of Methylene, Cyclopentadienylene and Diphenylmethylene," *J. Am. Chem. Soc.*, **81**, 5874 (1959).
203. Frey, H. M., "The Photolysis of Diazoethane and the Reactions of Ethylidene," *J. Chem. Soc.*, 2293 (1962).
204. Murray, R. W., and A. M. Trozzolo, "Dicarbenes. The Preparation and Some Reactions of 1,4-Bis(α-diazobenzyl)benzene," *J. Org. Chem.*, **26**, 3109 (1961); A. M. Trozzolo, R. W. Murray, G. Smolinsky, W. A. Yager, and E. Wasserman, "The E.p.r. of Dicarbene and Dinitrene Derivatives," *J. Am. Chem. Soc.*, **85**, 2526 (1963).
205. Kistiakowsky, G. B., and B. H. Mahan, "The Photolysis of Methyl Ketene," *J. Am. Chem. Soc.*, **79**, 2412 (1957).
206. Holroyd, R. A., and F. E. Blacet, "The Photolysis of Dimethyl Ketene Vapor," *J. Am. Chem. Soc.*, **79**, 4830 (1957).
207. Cairns, T. L., *et al.*, "Cyanocarbon Chemistry. I. Preparation and Reactions of Tetracyanoethylene," *J. Am. Chem. Soc.*, **80**, 2775 (1958).

208. DOERING, W. v. E., and P. M. LAFLAMME, "A Two-step Synthesis of Allenes from Olefins," *Tetrahedron,* **2,** 75 (1958).

209. SKATTEBØL, L., "Allenes from *gem*-Dihalocyclopropane Derivatives and Alkyllithium," *Tetrahedron Letters,* 167 (1961).

210. MOORE, W. R., and H. R. WARD, "Reactions of *gem*-Dibromocyclopropanes with Alkyllithium Reagents. Formation of Allenes, Spiropentanes, and a Derivative of Bicyclopropylidene," *J. Org. Chem.,* **25,** 2073 (1960); W. R. MOORE, H. R. WARD, and R. F. MERRITT, "The Formation of Highly-strained Systems by the Intramolecular Insertion of a Cyclopropylidene: Tricyclo-[4.1.0.0^{2,7}]heptane and Tricyclo[4.1.0.0^{3,7}]heptane," *J. Am. Chem. Soc.,* **83,** 2019 (1961).

211. BAWN, C. E. H., and W. J. DUNNING, "Chemiluminescence of Sodium Vapour with Organic Halides," *Trans. Faraday Soc.,* **35,** 185 (1939); C. E. H. BAWN and J. MILSTED, "The Stability of Hydrocarbon Biradicals and their Reactions," *Trans. Faraday Soc.,* **35,** 889 (1939).

212. FRANZEN, V., and H.-I. JOSCHEK, "Untersuchungen über Carbene. VI. Thermische Spaltung von Tetra-α-naphthyl-äthylen in Di-α-naphthylmethylene," *Ann.,* **633,** 7 (1960).

213. WANZLICK, H.-W., "Aspects of Nucleophilic Carbene Chemistry," *Angew. Chem. Intern. Ed. Engl.,* **1,** 75 (1962).

214. BRESLOW, R., "On the Mechanism of Thiamine Action. IV. Evidence from Studies on Model Systems," *J. Am. Chem. Soc.,* **80,** 3719 (1958).

215. CLOSS, G. L., "Carbenes from Alkyl Halides and Organolithium Compounds. IV. Formation of Alkylcarbenes from Methylene Chloride and Alkyllithium Compounds," *J. Am. Chem. Soc.,* **84,** 809 (1962).

216. BRINTON, R. K., "The Photolysis of Acetaldazine," *J. Am. Chem. Soc.,* **77,** 842 (1955).

217. ZIMMERMAN, H. E., and S. SOMASEKHARA, "The Mechanism of the Thermal Decomposition Reaction of Azines," *J. Am. Chem. Soc.,* **82,** 5865 (1960).

218. OKABE, H., and J. R. MCNESBY, "Vacuum Ultraviolet Photolysis of Ethane: Molecular Detachment of Hydrogen," *J. Chem. Phys.,* **34,** 668 (1961).

Author Index

Subject Index